Jacqueline Pacifica Oliveira de Sa

Characterization and serological survey of the Influenza Virus

Jacqueline Pacifica Oliveira de Sa

Characterization and serological survey of the Influenza Virus

Etiology of Acute Respiratory Infections

ScienciaScripts

Imprint

Any brand names and product names mentioned in this book are subject to trademark, brand or patent protection and are trademarks or registered trademarks of their respective holders. The use of brand names, product names, common names, trade names, product descriptions etc. even without a particular marking in this work is in no way to be construed to mean that such names may be regarded as unrestricted in respect of trademark and brand protection legislation and could thus be used by anyone.

Cover image: www.ingimage.com

This book is a translation from the original published under ISBN 978-613-9-61516-2.

Publisher:
Sciencia Scripts
is a trademark of
Dodo Books Indian Ocean Ltd. and OmniScriptum S.R.L publishing group

120 High Road, East Finchley, London, N2 9ED, United Kingdom
Str. Armeneasca 28/1, office 1, Chisinau MD-2012, Republic of Moldova, Europe
Printed at: see last page
ISBN: 978-620-7-85615-2

Table of contents:

Chapter 1   3

Chapter 2   22

Chapter 3   26

Chapter 4   29

Chapter 5   36

Chapter 6   42

Chapter 7   46

Chapter 8   48

Chapter 9   48

Chapter 10   53

PREFACE

Respiratory diseases are the main cause of human morbidity and occupy a prominent place in health statistics.

They cause absenteeism from school and work, in the latter case implying a serious economic problem.

On the other hand, they have a varied symptomatology, with different syndromic aspects ranging from upper respiratory tract manifestations (coryza, pharyngitis) and laryngo-tracheal involvement (croup), to the involvement of the lower pores (bronchi and lungs), with the production of severe cases of bronchiolitis and pneumonia.

These manifestations can be caused by a wide variety of microorganisms (bacteria, fungi, protozoa), alongside viruses.

It should be noted that the respiratory syndromes mentioned above do not have specific characteristics, due to the nature of the etiological agent, making it almost impossible to clinically infer the causal element.

The list of viruses that can cause respiratory infections is quite extensive, including Orthomixovirus (Influenza), Parainfluenza, and Respiratory Syncytial Virus, Picorvavirus (Enterovirus - Coxsackie and Echo; Rhinovirus) Coronavirus, Adenovirus and Reovirus.

As you can see, about half of the major groups of viruses recognized today can be located in the respiratory tract, causing manifestations, such is the breadth of the spectrum of agents to be taken into account.

The lack of knowledge about the role of the Influenza virus in our environment led us to carry out this work, which has already resulted in a monograph and, together with other respiratory viruses, a master's dissertation.

It is well known that, in temperate countries, these agents play an important role in infections, and the desire to develop a study that can help the scientific community and the population led us to carry out this work.

We believe that, given the scarcity or even absence of information on the subject in our field, the elements we can gather will make an interesting contribution.

In this way, public health specialists will be able to count on basic data for future studies that, to a certain extent, clarify the situation in our state with regard to the course of the Influenza virus in the population studied here, over 65 years of age, to be vaccinated in order to determine the immunogenicity profile of a vaccine containing new strains in Maceió/AL.

Chapter 1
l.Introduction
1.1 Acute respiratory infections a global problem

Among infectious and contagious diseases, acute respiratory infections (ARIs) are the most common worldwide. Their prevalence and incidence on all continents, affecting all age groups, from newborns to the elderly, and the multiplicity of their etiological agents, make ARIs undoubtedly one of the world's biggest public health problems. In addition, its clinical spectrum, ranging from a simple cold to viral pneumonia, requires an understanding of its epidemiological and pathophysiological characteristics.

ARI conditions classified as upper airway (HIGH ARI), such as the common cold, are frequent, generally self-limiting and rarely with complications that require clinical assessment and treatment. They affect all age groups, but are more frequent in children. Studies carried out in closed communities in the USA show that 6 to 8 episodes of ARI occur per year in every child aged 0 to 2 years. Analysis of the entire American population reveals an incidence of 0.8 episodes/person/year of upper airway ARI. These infections are a nuisance for school-age children, causing absences from school and a large number of visits to health services. For economically active adults, ARIs are generally the most frequent cause of absence from work, thus causing damage to services.

In the elderly, as in children, ARIs are once again a major reason for seeking medical attention, as well as being a serious problem for debilitated patients or those living in institutions.

ARIs have the greatest impact on children, where they involve the lower respiratory tract (LUTS) and can be a cause of death. This is more relevant in developing countries, where it is estimated that ARIs are responsible for a third of deaths in children up to 5 years of age, and it is estimated that every 7 seconds a 5-year-old child dies from ARI. In 1977, Bulha and Hitze presented data on ARI mortality from 88 countries affiliated to the World Health Organization (WHO). Based on figures from the period 19701973, the rate of 666,000 deaths/year at that time was presented. It was then estimated that the total number of deaths from ARI would be around 2.2 million a year, since the aforementioned data referred to A of the world's population. Mortality from ARI in developing countries is much higher than in developed countries, even considering that many locations do not have population data. More specific figures show ARI mortality rates of 1500/100,000 live births in countries such as Egypt (1977), Paraguay (1977) and Mexico (1976), and these rates are around 30 times higher than those observed in Canada and the USA (Chretien et al, 1980).

Based on demographic statistics, there are currently 4.5 million deaths from ARI in children under the age of five, the majority of which occur in developing countries.

Chretien et al. (1980) demonstrated the magnitude of the ARI problem in developing countries by looking at mortality rates from other causes, comparing developed and developing countries. It can be seen that the mortality rate from ARI in developing countries is higher than the rates from lung cancer and cardiovascular diseases, while in developed countries the exact opposite is true. Pio and collaborators from the WHO (see Cintra, 1997), showed that the total incidence of ARI does not differ between developed and developing countries, but there are marked differences in the mortality rate from ARI, which can be attributed to differences in the severity of the diseases, the costs of care and the adequacy of treatment, especially in children and the elderly.

Since 1977, the WHO has been developing programs in the area of respiratory infections, within the context of technical cooperation between developing countries for the control of ARIs. Initially, efforts are focused on characterizing the problem in terms of morbidity and mortality and identifying the population most at risk. This includes etiological diagnosis, determining the importance of each agent, as well as the main signs, pathophysiological and epidemiological characteristics in the different regions of the world. Based on the aforementioned data, it becomes possible to plan control measures, geared to the characteristics of each location.

1.2 Etiology of Acute Respiratory Infections

Until the mid-1950s, little was known about the etiology of ARIs. Bacteria such as streptococci, hemophilus, staphylococci and influenza A viruses were known, as were the so-called filterable agents. As microbiological techniques improved, new agents began to be identified, and those already known could be better characterized.

In the 1950s and 1960s, studies on the epidemiology of ARIs began to emerge in field experiments in research communities, identifying the main etiological agents, be they bacteria, viruses, fungi or protozoa.

Respiratory diseases account for 95% of all illnesses that affect humans, and those of viral etiology are responsible for more than 90% of acute respiratory diseases (ARIs), making them among the most studied today (Tarantino, 1990). The common cold, of unclear etiology, contributes significantly to ARDs, with between 5 and 7 episodes of flu-like ARIs.

Viruses are the most frequent agents of acute respiratory infections worldwide, with more than two hundred viruses identified as causing respiratory conditions ranging from the common cold to potentially life-threatening pneumonia. Since the discovery of the Influenza A virus in 1933, much has been studied about these agents and knowledge about respiratory viruses has undoubtedly increased greatly. This has been most evident in the last two decades, with the advent of faster techniques for virological diagnosis, making it increasingly common to identify these agents, which previously depended on more laborious and time-consuming processes, such as viral isolation in cell culture, which favors epidemiological studies.

Rhinoviruses are the most frequent agents of the common cold, while Parainfluenza and Respiratory Syncytial Virus (RSV) are the most frequent agents of pneumonia in children. The Parainfluenza 1 virus is the most frequent agent of laryngotracheobronchitis (croup) and the Influenza virus is the agent of influenza, with characteristics of worldwide spread and the occurrence of pandemics (see tab. 1.1).

Tab.1.0 Main types of virus associated with clinical syndromes of the respiratory tract

Virus	Types involved	Sick	Nucleic acid

Rhinovirus (>100 types)	several at any given time in the community	Common cold	Negative single-stranded RNA
Coxsackie A virus (24 types)	Especially A21	Common cold; also oropharyngeal vesicles	
Influenza Virus	A, B and C	can also reach the lower respiratory tract	Negative single-stranded RNA
Parainfluenza viruses (4 types)	1,2,3 e 4	It can also affect the larynx	Negative single-stranded RNA
Respiratory syncytial virus	1 single type	It can also reach the lower respiratory tract	Negative single-stranded RNA
Coronavirus (various types)	all	Common cold	Negative single-stranded RNA
Adenovirua(41 types)	5-10 types	Pharyngitis and conjunctivitis	Negative double-stranded adenovirus

Echovirus	11,20	Common cold	

Jawetz, et all 1991.

Alterations in nonspecific defense mechanisms during viral airway infections facilitate the penetration of bacterial agents that cause secondary infections, especially in malnourished children.

Therefore, it is essential to constantly monitor the etiology and epidemiology of ARIs around the world, as well as to analyze the antigenic variants of these agents, especially viruses. These variants represent difficulties in the acquisition of immunity by humans. Molecular biology studies of viral agents are the current alternative for better characterization of these variants, thus allowing the development of effective vaccines against the main agents of ARIs.

1.3 Influenza virus

Epidemics have caused problems in different populations around the world, including a high frequency of morbidity in different sectors: schools, public services, nurseries and other activities. In developed countries, influenza is an important cause of both morbidity and mortality. One of the age groups most quickly affected and with the highest mortality rate during an influenza epidemic is the elderly.

Experts indicate that the more developed countries are, the greater the impact of influenza epidemics. This impact is also seen in developing countries, varying in intensity from individual to individual and from country to country. Among the viral agents of ARIs, influenza causes acute respiratory infections that are generally self-limiting and occur in outbreaks of varying severity almost every winter.

The term Influenza has Italian origins and dates back to the 14th and 15th centuries when, in the region of Florence, it was believed that the influences of unusual planetary conjunctions acted in the genesis of endemic periods of fever, cough and chills. Therefore, these epidemics were caused by the influences of the stars and the impact of an Influenza epidemic, with a sudden onset, caused rapid spread over a considerable part of the population, affecting children, young people and the elderly (Oliveira, 1994).

In 1933, the isolation of a virus by Smith at all established the infectious etiology of the disease, leading to the precise definition of its epidemiology and clinical variation. The virus isolated in 1933 is now known as the type A influenza virus.

A second type of virus was isolated in 1940, the type B influenza virus. In 1950, type C influenza was isolated from a patient in non-epidemic circumstances by Taylor and was later shown by Francis and collaborators (1957) to be the cause of the epidemic disease.

The severity of respiratory infections varies, from a simple cold to a flu-like process with fever and prostration, or severe bronchitis and bronchopneumonia. Although the pharynx, larynx and upper respiratory tract are the most frequently affected, the infection can also affect the conjunctiva and salivary glands. For the most part, these diseases are highly contagious, and many usually occur in the form of epidemics. Our knowledge of their etiology is far from complete, but myxoviruses are responsible for a considerable proportion of respiratory tract diseases.

Myxoviruses cause respiratory infections in humans and animals. They owe their name to their affinity for the mucus in the respiratory tract and other regions of the body. They vary in size from 60-200Dm (usually 80-150Dm) and are spherical or filamentous in shape. They contain a central mass of RNA with protein capsomeres arranged along their spheres. They also have a lipid coating membrane, which makes them sensitive to 20% ether. Many components of the myxovirus group, although not all, possess an enzyme that causes erythrocytes to agglutinate (viral hemagglutination), whereby the viruses bind directly to erythrocytes (red blood cells) resulting in the visible clumping of these cells. The ability to hemagglutinate is restricted to certain viruses, and for those that possess it, the species of animal host is important. Viral hemagglutination is used in virology laboratories to detect hemagglutinating viruses such as influenza.

There are two subgroups of Mixovirus known as Orthomixovirus and Parainfluenza (Waterson, 1962). The first subgroup includes influenza viruses A, B and C and influenza viruses of pigs, ducks, horses and birds. The second subgroup includes epidemic parotitis virus, Newcastle virus and Parainfluenza 1,2,3 and 4.

Myxoviruses have, to varying degrees, the ability to remove or destroy receptors on the surface of red blood cells. The agent that destroys the cell receptor is enzymatic in nature, and this receptor consists of a mucopolysaccharide containing neuraminic acid, which is specifically attacked by the virus enzyme, neuraminidase (Waterson, 1962).

Influenza A virus represents the main etiologic agent of respiratory diseases in the USA and in most other parts of the world. It causes a life-threatening lower respiratory tract ARI in immunosuppressed adults (Kilbourne, 1992). Influenza causes typical epidemic outbreaks and pandemics around the world, annually resulting in approximately 30,000 deaths in the USA (Lui and Kendal, 1985).

1.4 The respiratory tract as a gateway

Air normally contains suspended particles, including smoke, dust and microorganisms. Thanks to the existence of approximately 500-1000 microorganisms per m^3 inside buildings and a ventilation rate of 6 liters per minute at rest, around 10,000 microorganisms are introduced into the lungs. In the lower or upper respiratory tract, inhaled microorganisms and other particles are caught in the mucus, transported to the back of the throat by ciliary action and finally swallowed. Those that invade the normal, healthy respiratory tract manage to escape these mechanisms because they have developed specific systems for invading and attaching to the surfaces of the cells that form the mucociliary layer.

Table 1.1 Some attachment molecules, usually called adhesins, from different types of viruses and related organisms.

Tab. 1.1 Microbial attachment to the respiratory tract

Microorganism	Sick	Microbial adhesin
Influenza Virus	Influenza	hemagglutinin
Rhinovirus	Common cold	protein capsid
Parainfluenza type 1	respiratory disease	protein envelope
VRS	respiratory disease	protein envelope
Mycoplasma pneumoniae	Atypical pneumonia	mycoplasmic molecule
Adenovirus	respiratory disease	protein capsid
Haemophylus influenzae	respiratory disease	Surface molecule

Jawetz, et all 1991.

Efficient cleaning or purification mechanisms control these constantly inhaled particles and another way of interfering with purification mechanisms is to inhibit ciliary activity. This facilitates the establishment of microorganisms in the respiratory tract. Inhaled organisms that reach the alveoli encounter alveolar macrophages, whose function is to remove foreign particles and keep the air spaces clean. Most microorganisms are destroyed by these macrophages and some escape phagocytosis, thus avoiding destruction. Alveolar macrophages are damaged after inhaling toxic particles from asbestos and certain dusts, which can lead to increased susceptibility to respiratory pathogens.

Viruses are the most common invaders of the nasopharynx, and although there are a wide variety of types that cause respiratory infection, two types are restricted to the surface and spread throughout the body (tab. 1.3).

Tab. 1.2 Types of respiratory infection

Types	Examples	Consequences
Restricted to the surface	Influenza, Adenovirus, RSV	Local spread, local defense of mucous membranes
Spread throughout the body	Measles, Mumps, Rubella	Little or no penetration site, being through the body

Adapted from Oliveira, 1994

Rhinoviruses and Coronaviruses together cause more than 50% of common colds. These viruses induce a flow of fluid rich in viral particles and when the sneeze reflex is triggered, a large number of viral particles are eliminated into the air. Transmission therefore takes place via the aerogenous route and also via contaminated hands. Most of these viruses have surface molecules that allow them to attach to the surface of the host's cells, which may contain cilia or microvilli protruding from them. As a result, they are not eliminated in the secretions and are able to initiate infection in the normal individual. From the first infected cells, the viral offspring spread to neighboring cells and, through the surface secretions, reach other sites on the mucosal surfaces. After a few days, damage to the epithelial cells and the secretion of fluid containing inflammatory mediators such as bradykinin lead to the typical symptoms of the common cold (Waterson, 1962).

1.5 Properties and characteristics of Orthomyxoviruses

All known Orthomixoviruses are called Influenza. There are three known immunological types (because they confer different immunities): A, B and C. Antigenic changes occur continuously in group A influenza viruses, and to a lesser extent in type B, while type C appears to be antigenically stable.

Influenza viral particles are generally spherical and about 80-120^m in diameter, although virions can vary greatly in size. Long filamentous forms (up to many micrometers in length) are commonly observed during the initial passages of the newly isolated virus.

The single-stranded RNA genomes of the Influenza A and B viruses are divided into eight separate segments. Influenza viral particles contain 7 different structural proteins, 3 of which are proteins (PB 1, PB 2 and PA) that are bound to the viral RNA and are responsible for RNA transcription and duplication. The nucleoprotein (NP) binds to the viral RNA, forming a 9Dm diameter structure with a helical configuration. The matrix protein, designated M because of its association with the viral membrane (Waterson, 1962).

These glycoproteins encoded by the virus are Hemagglutinin (HA) and Neuraminidase (NA), which are integrated into the envelope and exposed as molecules approximately 10 Dm long on the surface of the particle. These two surface glycoproteins are important antigens that determine the antigenic variation of influenza viruses. Hemagglutinin represents around 25% of the viral protein.

Influenza viruses are relatively resistant and can be stored at 0 to 4°C for weeks without losing their viability. The virus loses its infectivity more quickly at -20°C than at 4°C. Ether and protein denaturants decrease infectivity. Hemagglutinin and internal antigens are more stable to inactivation than the viral particle. Both infectivity and hemagglutination are more resistant to inactivation at alkaline pH than at acidic pH.

The influenza virus has a core, which is ribonucleic acid, divided into independent pegs, the genome. There is a polymerase reverse transcriptase involved in the replication of the viral genome, and a basic protein and a lipid layer form the lipid envelope from the cell surrounding the viral particle. In the envelope there are spicules made up of glycoproteins of complex antigenicity which are absolutely distinct from the other constituents.

Variations in these two antigens are responsible for the occurrence of epidemics and pandemics. Knowledge of this variation makes it possible to predict the severity of the infections caused by the viruses at an early stage.

Orthomyxoviruses have biological properties such as: cell fusion (neuraminidase) and hemadsorption (hemagglutinin) similar to the Parainfluenza virus (Collie and Oxford, 1993).

1.6 Classification and nomenclature

The internal structural proteins, NP and M, are used for the replication of influenza virus types A, B and C. These proteins are not cross-reactive between the three types. Antigenic variations in the surface glycoproteins (HA and NA) are used to subdivide the viruses.

The standardized naming system for isolated influenza viruses includes the following data: type, host of origin, geographical origin, strain number and year of isolation. The antigenic descriptions of HA and NA are given in brackets for type A viruses. The host of origin is not indicated in the case of viruses isolated in humans, e.g. A/Hong Kong/03/68(H3N2), but it is for other animals, e.g. A/pig/Iowa/15/30(H1N1) . Thus, Influenza A/Shanghai/11/87(H3N2) which indicates an influenza virus of type A and subtype (H3N2), isolated in 1987 in Shanghai-China.

As for type B of the influenza virus, we don't find subtypes but only its characterized strain, for example B/Beijing/11/90.

1.7 Structure and function of Hemagglutinin (HA)

Hemagglutinin is responsible for the penetration of the virus into the cell via the receptor. Antibodies against this protein neutralize viral infectivity and are thus the main determinant of humoral immunity. Viral Neuraminidase helps to release the virus into the cell, anti-neuraminidase antibodies are not neutralizing, but they limit viral reproduction and thus the intensity of infection.

Influenza B and C viruses have been less studied, but appear to be structurally similar to Influenza A viruses. B and C viruses are mainly associated with sporadic epidemics in children and adults, in schools or other populations kept in similar establishments. Most adults carry antibodies against these viruses, probably as a result of recurrent subclinical infection, and there are no indications that pandemics of Influenza B or C have occurred.

Discussions around influenza generally refer to type A influenza viruses, as these are the most important viruses because they are the most mutable. One of the regular and most notable characteristics of influenza is the frequency with which changes in antigenicity occur, which is more frequent with influenza A than with influenza B and C. These changes help explain why influenza continues to be responsible for annual inter-epidemic outbreaks, epidemics and pandemics. Epidemics occur every 2-4 years and pandemics occur every decade. These epidemics occur almost exclusively during the winter months - October to April in the northern hemisphere and May to September in the southern hemisphere, and simultaneous outbreaks of Influenza A and B, or Influenza A and RSV (respiratory syncytial virus) infection have been demonstrated.

Studies indicate that the strains circulating at the end of a seasonal epidemic are likely to cause an outbreak in the following season. Influenza pandemics, on the other hand, result from the emergence of a virus to which the entire population has no immunity, so that the influenza epidemic progresses and involves all regions of the planet. The most serious pandemics have occurred when there have been significant antigenic changes in the two main surface antigens (Hemagglutinin and Neuraminidase).

The HA protein of the influenza virus has been studied in detail because of its biological importance. This protein binds viral particles to susceptible cells and is the main antigen against which neutralizing (protective) antibodies are directed. Its variability, which consists of changes in the sequence of its amino acids, is the main characteristic responsible for the continuous evolution of new strains and subsequent flu epidemics. HA derives its name from its ability to agglutinate erythrocytes under certain conditions. The HA spike of the viral particle is a trimer made up of 3 intertwined HA1 and HA2 dimers. This trimerization gives the spike greater stability than could be achieved with an HA monomer, and we have found the structural characteristics of the influenza virus. The Hemagglutinin (HA) and Neuraminidase (NA) spicules are projected from the bilipid layer; underneath this layer the matrix protein (M), surrounds the segmented RNA genome and each segment is surrounded by the nucleocapsid protein.

Fig. 1.0 With regard to the architecture of influenza viruses, examining a virus particle as seen in an electron microscope reveals its complex structure.

Fig 1.0 Micrograph and schematic structure of the Influenza virus, according to Kaplan & Webster,

1 - Hemagglutinin, 2- Neuramin, 3- Protein matrix, 4- Lipid layer, 5- Polymerase, 6- Nucleoprotein, 7- RNA.

The cellular receptor (viral binding site) is located at the top of each large HA glycoprotein globule (fig. 1.4 below) where we find the influenza virus with its HA and NA surface glycoproteins.

(A) Primary structure of HA and NA (polypeptides). The cleavage of HA into HA1 and HA2 is necessary for the virus to become infectious. HA1 and HA2 remain linked by a disulfide bridge(S-S)No post-translational cleavage occurs in neuraminidase. The binding sites (carbohydrates) are shown. The hydrophobic amino acids that hold the proteins in the viral membrane are located near the C-terminus of HA and the N-terminus of NA.

(B) Packaging of HA1 and HA2 polypeptides in an HA monomer.

(C) Carboxyterminal residues are projected through the membrane.

(D) Structure of the NA tetramer. Each NA molecule has an active site on its upper surface.

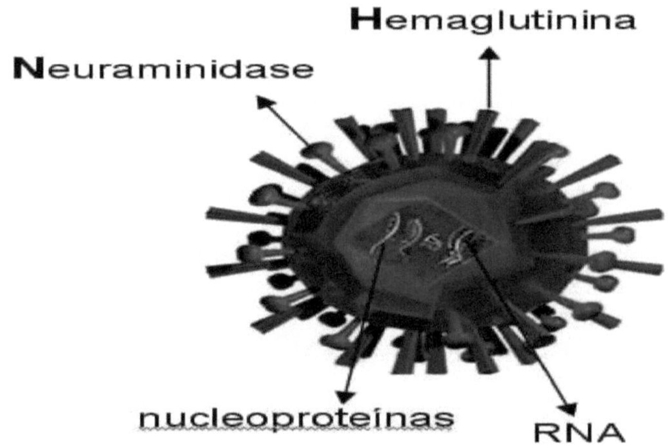

Fig. 1.0 Structure of Hemagglutinin and Neuraminidase

The HA spike of the viral particle is a trimer made up of 3 intertwined HA1 and HA2 dimers (fig. 1.4). This trimerization gives the spike greater stability than could be achieved with an HA monomer. The cell receptor (viral binding site) is located at the top of each large globule of the HA gliocoprotein.

Fig 1.0 Influenza antigens: HA and NA, the triangular-shaped HA and mushroom-shaped NA protrude from the viral lipid layer in a 5:1 ratio (HA;NA). The Matrix protein forms a layer below the lipid membrane and surrounds the eight viral RNA segments.

1.8 Structure and function of Neuraminidase (NA) (fig 1.4)

The antigenicity of neuraminidase, which is another gliocoprotein on the surface of influenza viral particles, is also important in determining the subtype of the virus isolated.

The complete sequence of neuraminidase is known. The spike in the viral particle is a tetramer made up of 4 identical monomers. A thin stalk has a box-shaped structure at the top. There is a neuraminidase catalytic site at the top of each head, so each neuraminidase spike contains 4 active sites.

Neuraminidase appears to act at the end of the virus's life cycle. It facilitates their release from infected cell surfaces during the budding process and helps prevent the virions from self-aggregating. The cellular receptor (viral binding site) is located at the top of each large HA glycoprotein globule (fig. 1.4) where we find the influenza virus with its HA and NA surface glycoproteins.

1.9 Evolution of HA and NA in epidemic variants and selection of viruses for vaccines

The remarkable ability of the human influenza virus to cause successive infections over a lifetime is due primarily to the ability of the HA protein to greatly modify its sequence without jeopardizing its essential function. This sequence variation is evidenced by the antigenic variation which belongs to a new variant capable of evading the immunity previously acquired by the host through infection or vaccination with previous variants. Determining the molecular basis of this antigenic variation has been essential in understanding how the HA molecule is involved in the immune response (Cox, 1993).

The two surface antigens of influenza viruses undergo antigenic variations independent of each other. These antigenic variations involve the HA and NA proteins present on the viral envelope, and these variations are one of the main, if not the main, obstacle to the development of an effective flu vaccine. These antigenic variations are called Antigenic Mutations (Shift) and Antigenic Trends (Drift).

The antigenic mutation occurs in the A virus, but not in the B or C viruses. There is a sudden rearrangement of genomes, since a single cell can simultaneously have two different viruses, which will cause a mixture of RNA to be duplicated and expressed in viral replication. As influenza also affects other animals such as birds, this rearrangement can occur between humans and birds, resulting in a new and distinct subtype of influenza. These antigenic mutations, which result in large genetic variations in HA and NA, probably arise from recombination between human and avian Influenza A strains.

Antigenic tendency is described as a rapid change and therefore more frequent than the Shift, because it is at the level of protein synthesis, i.e. a change in the amino acid sequence. These small antigenic variations result from mutations in the genes that code for HA and NA. This means that the new viral samples tend to differ from the original sequence.

We note that the internal proteins of viruses, such as the nucleoprotein (NP), do not undergo these genetic alterations (Stokes, 1988).

1.10 Influenza A(H1N1) virus HA evolution

The Influenza A(H1N1) virus has circulated during two distinct periods in this century. The first period was approximately 1919-1957, while the second began in 1977 and continues to circulate today. Since 1977, the A(H1N1) virus has caused illness primarily in children and young adults, and has been widely spread by individuals who were exposed to this strain, except for individuals who were exposed to similar viruses that circulated during the early 1950s. Sequence analysis of circulating H1N1 strains isolated from 1950-1957 and 1977-1983 revealed a remarkable genetic similarity between the circulating HA genes of A/Fort Warren/50 and A/USSR/77 and established two distinct evolutionary pathways for the HA of the H1N1 virus circulating during the two separate periods.

The sequence change occurring in the HA gene domain of the contemporary H1N1 strain continues to be monitored. Separate strains of the H1N1 virus can co-circulate and the epidemic strain of A/Taiwan/86 does not evolve directly from the A/Chile/83 strain. Co-circulating strains can complicate the process of selecting vaccine strains, as was evident during the unexpected emergence of the A/Taiwan/86 virus when a vaccine based on the A/Chile/86 strain had been prepared. This led to the manufacture and distribution in 1986 of a supplementary vaccine containing the A/Taiwan/86 strain.

Since the emergence of the A/Taiwan/86 variant, small antigenic variations have been detected among the H1N1 viruses circulating in the world. Monitoring the sequence change in the HA of circulating strains and isolates during these periods indicated that the HA of this subtype has continued to evolve at a steady rate.

A value of 3.59×10^{-3} nucleotide substitutions per site per year was obtained for the HAs of viruses isolated between 1977 and 1992, and a similar value of 3.7×10^{-3} nucleotide substitutions per site per year was obtained for the HA of 20 viruses isolated between 1986 and the present. The amino acid substitution type was calculated by similar methods to be 0.58 per year for the 46 strains isolated between 1977 and 1992. These values are somewhat lower than those previously reported for three reasons: 1) different methods were used to calculate them, 2) nucleotide and amino acid exchange due to host-mediated variations and 3) a large number of circulating strains isolated.

1.10.1 Influenza A(H3N2) virus HA evolution

Since its emergence during the 1969 pandemic, the Influenza A(H3N2) virus has been found continuously in humans and has caused morbidity and mortality during successive epidemics, sequence data are now viable for multiple viruses in the recent influenza season and with a wide geographical distribution. Using this data we have observed that early before the circulation of a new epidemic variant, there can be a small variation in amino acid level sequence between distinct viruses isolated geographically over a period of nine months. The variation that is observed for these 1987 reference strains is likely to be host-mediated and the amino acid change that was observed may assess the antigenic heterogeneity observed between these strains. This observation indicates that although considerable sequence similarity exists between epidemic variants, an unexpected sequence may emerge and be present in another epidemic.

This A(H3N2) subtype has generally been associated with excess mortality in the elderly and patients at high risk of influenza-related complications, matching this vaccine component to the circulating strain is particularly important. Thirteen of the fourteen different inactivated strains of the (H3N2) subtype have been used as components of the influenza vaccine since 1968.

1.11 Influenza type B virus evolution

The Influenza B virus also causes epidemic periods of illness with each epidemic usually being associated with the emergence of a new antigenic variant. Influenza B epidemics occur most frequently in schools, but epidemics have also been described in nursing homes, indicating that Influenza B can affect all age groups.

The HA genes of viruses representing each of the two lineages from a variety of geographical locations have been sequenced to gain a better understanding of the evolution of Influenza B HA. Two antigenically distinct strains of Influenza B, one represented by B/Victoria/2/87 and the other represented by B/Yamagota, have been co-circulating and causing disease in humans since at least 1983. Separate strains of Influenza B appear to have a longer circulation than Influenza A.

The identification methods used to calculate the change types for Influenza A HA were also used to calculate the change types for Influenza B. With a nucleotide substitution rate of 1.96 x 10^{-3} per site per year and amino acid changes of 0.30% per year, this result was extracted from the 50 circulating strains isolated between 1979 and 1992 (Cox, 1993).

1.12 Why do new influenza pandemics always start in South Asia?

The fact that influenza pandemics have always started in South Asia in these centuries of pandemics is due to the fact that it is a region with many closed pig farms and open duck farms, causing infections between them. In this region there are two problems: the infection of pigs, humans and birds. This would cause a situation of genetic recombination due to the co-habitation of humans and pigs and this interaction of the people who work with the birds are basically the same as those who work with the pigs, causing a very strong exchange of organic material (feces), triggering new pandemics.

So these new Influenza A pandemics are characterized by the crossing of bird, pig and human viruses, with the human virus chain/sequence prevailing. Some human viruses are disseminated by birds and vice versa, assuming a continuous course of exchanges between species. Pigs, on the other hand, show a low prevalence/recombination of human and bird viruses. It was possibly this exchange that caused/originated the 1918/1919 pandemics (Ludwig and Haustein, 1992).

In this case, when an important antigenic change occurs in the virus, it can be so extreme "SHIFT" that the immunity acquired by the population to previously resistant Influenza A viruses is inadequate to prevent infection or disease initiated by the new virus, which replaces the old virus completely. Under these conditions, people of all ages and in all places are susceptible to influenza and a worldwide epidemic or pandemic can occur, as in 1918 with Influenza A(H1N1) and in 1957 with Influenza A(H2N2), both of which died out but both of which caused serious pandemics.

Influenza pandemics result from the appearance of a new virus against which the population has no immunity, so that influenza epidemics progress until they involve all parts of the world. The most intense pandemics occur when there are major antigenic changes in the main surface antigens HA and NA. An exception was Influenza A/USSR/77(H1N1), which did not cause an intense pandemic in 1977 to 1978, despite the antigenic mutation that occurred in both surface glycoproteins. This discrepancy perhaps reflects the fact that a large part of the world's population in 1977 was alive during the previous 1950's H1N1 era and thus had immunity (Hegele, et al 1992).

In 1957 there was a pandemic that appeared with different characteristics from the samples known until then; it was a new sample and found the population without any immunity. In a closed community, at the Beijamim Constant Institute (an institution for the blind), 93.8% of the cases studied were reported to have been affected in about 10 days in September; then the curve fell and plateaued. These patients were studied not only in terms of isolation (tab. 1.4 below), in which there were positive cases, confirmed by virus identification, and others in which the proof was essentially serological, because the sample did not have a characteristic behavior. But there was what is known as a seroconversion. In the acute phase, the hemagglutination inhibitor activity was 1/10; in convalescence it was 1/60. This diagnostic test is of great significance.

Tab. 1.4- Isolation of influenza viruses and antibody testing in patients at the Benjamin Constant Institute.

		Virus isolation		HI of patient sera
Registration	Symptoms	Harvest date	Results	Titles
19	Yes	04/09/57	Positive	1/40
20	Yes	04/09/57	0	1/10
21	Yes	04/09/57	0	1/20
22	Yes	04/09/57	0	0
23	Yes	04/09/57	0	1/20
24	Yes	04/09/57	0	1/10
25	Yes	04/09/57	0	0
26	Yes	04/09/57	0	1/20
27	Yes	04/09/57	0	0

28	Yes	04/09/57	Positive	1/80
29	Yes	04/09/57	0	1/20
30	Yes	04/09/57	Positive	1/20
31	Yes	04/09/57	0	1/20
32	Yes	04/09/57	Positive	1/80
33	Yes	04/09/57	0	0
34	Yes	04/09/57	0	1/20
35	Yes	04/09/57	0	1/10
36	Yes	04/09/57	0	1/10

In another closed community, the Aeronautical Hospital of Campo dos Afonsos, many samples were isolated. You can see the positive cases, all clinically very characteristic, with positive isolation in the majority and seroconversion in an appreciable margin of cases (tab 1.5)(Bruno Lobo M. 1959). During this epidemic, two very interesting cases were sent to us. They were from two children who had initially presented with a typical flu-like illness. An influenza sample was isolated from one of them using a throat wash. This patient died of meningoencephalitis, which is an inflammation of the nervous parenchyma when limited to the encephalon, with a marked involvement of the meninges. This is due to direct viral invasion or hypersensitivity aroused by a virus or other foreign protein. The diagnosis was made through a lumbar puncture of the cerebrospinal fluid, which showed a slight elevation of proteins. The other case was of a patient in the same room, who was not tested for the virus, but presented the same condition with nervous progression, and died. And on that occasion there were forms of pneumonia caused by the Asia 57 virus, which evolved into pneumonia and bacterial forms. Similar cases have been found extremely frequently here, in Chile and in the United States.

Tab 1.5- Influenza virus isolation and antibody testing in patients at the Campo dos Afonsos Aeronautics Hospital

		Virus isolation		HI of patient sera
Registration	Start of the start	Results	Collection after start-up	Titles
38	13/09/57	0	1 2	1/10 1/160
39	12/09/57	Positive	1 2	0 1/80
40	Uncollected material		1 2	0 1/80
41	14/09/57	Positive	1 2	1/80 1/20
50	18/09/57	0	1 2	0 1/20
77	15/09/57	0	1 2	0 1/10

From 1957 onwards, material began to be collected in order to monitor the samples, with a view above all to verifying their antigenic characteristics, and permanent isolations of the various respiratory viruses were made. All isolations carried out from 1957 onwards were studied; these samples were studied in terms of their antigenicity, using different isolation and identification techniques (Lobo, 1959).

Among the samples that have been isolated is Port Chalmers, which circulated in 1974, Scotland in 1975, Victoria in 1976, and A Victoria in 1977, which is related to England 42/72. In 1978, an outbreak of influenza was studied from April to June, in which a virus similar to England 47/72 was isolated, and in the same year a totally different sample from the last ones was isolated. This sample, called A1 Brazil/78, was similar to the one found in the 1946 pandemic (Melo, 1977). In 1980 and 81, other samples similar to the other A2H3N2 were isolated. Isolations from domestic and wild birds began in 1978 and 79. Although they were isolated at different times, they all showed the same HAV 6N3 immunological behavior (Chaves, et al 1979). The most interesting thing is that the virology department had been working with material from wild and domestic birds since 1972, having processed 2,513 materials for virus isolation, but it wasn't until 1978, confirming data from the literature(Bahl, 1975)(Becker, 1966)(Buttrdield, 1973), that the first bird viruses were isolated. Since then, the above viruses have continued to be systematically isolated from feces and not from water (Hinshaw, 1979), with the greatest number of isolations coinciding with the period from April to June, which is the same period in which the greatest number of samples of human origin have been isolated.

In a study carried out in the city of Maceió, the study population consisted of inpatients and outpatients with acute or chronic respiratory infections of the lower or upper respiratory tract, seen in the immunology/virology laboratory of the University Hospital (HU) of the Federal University of Alagoas, in the age groups: 0-4, 5-14, 15-24, 25-59 and >60 years of both sexes. The clinical specimen collected from the patients was serum up to the seventh day after the onset of symptoms. The Hemagglutination Inhibition (HI) technique (Palmer, et al 1975) was used to test for specific antibodies in the serum. The results were considered positive for dilution values above 1:40 and negative for values below and including 1:40. Five known subtypes of the influenza virus circulating in this population in 1996-97 were studied. Subtypes A/Taiwan/1/86(H1N1), A/Shangdong/9/93(H3N2) and A/Shanghai/11/87(H3N2) were found to be present in greater proportions in the study group, when compared to subtypes B/Panama 45/90 and B/Guangdong 8/93. In view of the results, it was concluded that the Influenza virus, which is one of the viruses that causes the most respiratory infections in the world, being responsible for outbreaks of epidemics and pandemics through the aforementioned subtypes, was present at high rates indicating wide dissemination in the population of Maceió/AL and that gender and age had no significant influence on the results(Cruz, M.E.M.; Zaidan, A.M.E.; Rodrigues, M.M.L.; Sá, J.P.O.; Silva, N.M.; & Oliveira, J.F. 1998).(tab 1.1) and (tab 1.2)

Tab. 1.1 Comparison of subtypes A/Taiwan/1/86(H_1N_1), A/Shangdong/9/93(H_1N_2), A/Shanghai/11/87(H_3N_2), B/Panama/45/20, B/Guangdong/8/93 in different age groups in the city of Maceió-AL.

Age group	Taiwan	Shangdong	Shanghai	Panama	Guangdong
0 - 4	73	79	78	25	19
5 - 14	100	95	87	35	0,8
15 - 24	100	93	74	82	33
25 - 59	100	81	75	56	26
>60	100	92	95	54	41

values expressed in %

Tab. 1.2 Comparison of subtypes A/Taiwan/1/86(H1N1), A/Shangdong/9/93(H3N2), A/Shanghai/11/87(H3N2), B/Panama/45/20, B/Guangdong/8/93 between the sexes in the city of Maceió-AL

Sex	Taiwan	Shangdong	Shanghai	Panama	Guangdong
Male	88	95	87	38	14
Female	98	92	78	58	28

Values expressed in %

1.13 Influenza virus replication
1.13.1 Virus binding, penetration and shedding:
The virus binds to the sialyl-oligosaccharide on the cell surface via the receptor located at the top of the large HA globule. The viral particles are then placed in endosomes. The next step is probably the fusion of the viral envelope with the cell membrane, triggering denudation. The amino terminus of HA, generated by the proteolytic cleavage of the HA precursor polypeptide, is essential for this stage. The low pH inside the endosome is ideal for virus-mediated membrane fusion. It is likely that this causes an alteration in the conformation of the hemagglutinin structure which brings the HA "fusion peptide" (fig. 1.4) into contact with the membrane. The nucleocapsids are then released into the cell cytoplasm.

1.13.2 Transcription and Translation:
The transcription mechanisms used by Orthomixoviruses differ greatly from those of other RNA viruses due to the fact that cellular activities are more closely involved. Viral transcription takes place in the nucleus. The polymerase encoded by the virus, consisting of a complex of three P proteins, is basically responsible for transcription. However, its action needs to be guided by the methylated and coated 5' ends, taken from newly synthesized cellular transcripts by cellular RNA-polymerase II. This explains why the duplication of the influenza virus is inhibited by dactinomycin and amanitin, which block cellular transcription, while other RNA viruses are not affected because they do not use cellular transcripts in the synthesis of viral RNA

(Jawetz, 1991).

1.13.3 Duplicate viral RNA:

The duplication of the viral genome is carried out by the same proteins (polymerase) encoded by the virus and involved in transcription. The mechanisms that regulate alternative duplication and transcription carried out by the same proteins are probably related to the abundance of one or more of the viral nucleocapsid proteins.

As with all segmented negative single strand viruses, they are called negative strand because the infective single strand does not encode proteins; instead, its complementary strand carries the coding sequences. Thus, the infective strand remains important in the absence of a preformed replicase (Alberts, et al 1994). In this case, the templates for viral RNA synthesis (transcription or duplication) remain inside the capsids. The only totally free RNAs are the mRNAs (messenger RNAs).

The first step in the duplication of the genome is the production of antigenic copies that differ from the mRNAs in both ends: the 5' ends are not coated and the 3' ends are not polyadenylated. These copies then serve as templates for the synthesis of faithful copies of the genome's RNAs.

There are common sequences at both ends of all viral RNA segments, which means that they can be efficiently recognized by the RNA synthesizing "machinery". It is not known how the segments are grouped together in the nucleocapsids of the virus progeny. The intertwining of genome segments from different "parents" in co-infected cells is probably responsible for the high frequency of genetic recombination typical of influenza viruses (Jawetz, 1991).

1.13.4 Maturabo:

The virus matures by budding through the apical surface of the cell. The individual viral components reach the budding site by different routes. The nucleocapsids are clustered in the nucleus, moving outwards from the cell surface. The glycoproteins, hemagglutinin and neuraminidase, are synthesized in the endoplasmic reticulum, modified and assembled into trimers and tetramers, respectively, and introduced into the plasma membrane.

The matrix protein (M), synthesized in the cytoplasm, serves as a bridge between the nucleocapsid and the cytoplasmic ends of the glycoproteins. The progeny of virions sprout out of the cell. Neuraminidase removes the terminal sialic acids from the glycoproteins on the cell surface, thus facilitating the release of viral particles from the cell and preventing their aggregation, so that each one can act as an isolated infecting unit. Otherwise, large clumps of particles would form due to the affinity of the hemagglutinin binding site for sialic acid. The viral multiplication cycle is rapid, and the new virus progeny is produced in 8 to 10 hours.

Many (90% or more) of the particles produced are not infectious. Sometimes the particles don't form capsids around the genome segments and often one of the large RNA segments is missing. These non-infectious particles can cause hemagglutination and interfere with the duplication of the intact virus. In addition to the typical spherical particles, elongated forms with the same surface glycoproteins can be produced, which also cause the agglutination of red blood cells characteristic of the virus (Jawetz, 1991).

Chapter 2
2. Pathogenesis and pathology

The influenza virus spreads from person to person via airborne droplets or contact with contaminated hands or surfaces. Some cells of the respiratory epithelium become infected when the viral particles are not removed by the cough reflex and escape neutralization by any pre-existing specific immunoglobulin A (IgA) antibodies or inactivation by unspecific inhibitors in the mucous secretions. New virions are then produced and spread to adjacent cells, where the duplication cycle is repeated. Viral neuraminidase reduces the viscosity of the mucus film in the respiratory tract, leaving the receptors on the cell surface exposed and promoting the spread of the virus-containing liquid to the lower portions of the tract. In a short time, many cells in the respiratory tract are infected and eventually destroyed.

The incubation period from the moment of exposure to the virus until the onset of the disease varies from 1 to 4 days, depending in part on the size of the viral inoculum and the immune conditions of the host. The spread of the virus starts the day before the onset of symptoms, reaches its maximum in 24 hours, remains high for 1 to 2 days, and then decreases rapidly. Interferon, which is a family of proteins induced by viral infection and blocks protein synthesis, is found in respiratory secretions around 24 hours after the virus begins to be eliminated. Influenza viruses are sensitive to the antiviral effects of interferon, and the interferon response is believed to contribute to the patient's recovery. Inflammation of the upper respiratory tract causes necrosis of the ciliated and goblet cells of the tracheal mucosa and bronchial mucosa, but does not affect the basal layer of the epithelium. Full regeneration of the cell damage probably takes a month. Interstitial pneumonia can be accompanied by necrosis of the bronchiolar epithelium and can be fatal. Epithelial damage to the respiratory tract by viruses reduces its resistance to secondary bacterial invaders, especially staphylococci, streptococci and Haemophilus influenzae.

It is likely that edema and infiltrates by mononuclear cells in response to cell death and shedding by viral duplication are responsible for the local symptoms, while the prominent symptoms associated with influenza virus infection are difficult to explain. Although the spread of the virus has been followed, the isolation of infectious viruses in the blood is very rare.

The virus primarily infects the respiratory epithelium and is determined by the inoculation of viruses from the respiratory secretions of infected individuals. Experimental studies have shown that a small number of inhaled viral particles, contained in a small aerosol particle with an average diameter of less than 10 Dm or in doses several times greater in a liquid suspension eliminated through the nose, produce the disease. Thus, infection can result from the transmission of infected secretions through interpersonal contact or fomites, and probably more frequently through the inhalation of aerosols from sneezing, coughing and other respiratory discharges of infected individuals.

Once the virus is deposited in the respiratory epithelium and unless it is prevented by specific secretory antibodies, mucoproteins or mechanical actions of the mucociliary mat, it attaches itself and penetrates the columnar epithelial cells by pinocytosis. Viral replication lasts 6-4 hours and the release of the virus continues for several hours before cell destruction ensues. The lining cells of the respiratory tract including the ciliated epithelium, alveolar cells, mucous gland cells and macrophages can be infected. The cytoplasmic changes are: granulations, vacuolization and swelling. Finally, the cells become necrotic and desquamate in some areas, being replaced by a smooth, metaplastic epithelium.

Secreted serum antibodies and immune mechanisms are not detectable at this stage, indicating that immunological mechanisms are probably not involved in fighting the disease, with the possible exception of circulating interferon which is detected during the acute stages of the disease, in respiratory secretions and in the serum of infected individuals, after one day of disease spread. The virus is sensitive to the antiviral effect of interferon, which contributes to the individual's recovery. Systemic symptoms and fever are due to the excretion of interferon, which suggests hematogenous spread of the virus or its release.

1 .1 Factors in the persistence of viral infections

For many years, doctors and researchers in virology have been paying attention to acute and febrile viral infections such as Polio, Influenza, Smallpox and Measles. There is a brief incubation period, followed by acute illness, which ends with recovery and presumably elimination of the virus. These important viral infections are gradually being controlled through vaccinations, and attention is turning to those viruses that persist in the body for months and years, despite the production of antibodies.

Persistent infections are a less important cause of acute diseases, but their persistence in human communities, their activation in patients undergoing immunosuppressive drugs and their association as a possible cause of immunological diseases or neoplasms, give them great interest today. One point of view on persistent infections refers to possible failures in the host's defense mechanisms, or methods by which microorganisms have overcome these defenses, and it seems interesting to consider persistent infections from this angle.

The host's defense mechanisms especially include the production of interferons, specific antibodies and specifically sensitized lymphocytes. Macrophages play an important defense role, particularly associated with cytolytic antibodies or immunologically generated lymphokines, and can be considered separately. Viruses that produce persistent infections tend to be those that cause a minimal cytopathological effect and this is an important factor in the persistence of infections and deserves special consideration. Viruses that cause severe destruction of the host cell probably induce an acute illness that ends with death or recovery. RNA viruses, on the other hand, show little or no cytopathic effect and cause persistent infections in their hosts. Viruses can persist despite destroying cells, if this occurs on a small scale and is limited by interferons. Presumably, there is a slow balance between the infecting viruses and the infected cells, and there is an integration with slow viral multiplication of a small number of viruses in a small amount of cells.

In persistent viral infections, there may be an ineffective immune response:
Tolerance: meaning a specific immunological hyporeactivity to the viral antigen, or:
1 - Autoimmunosuppression: this can be caused by the infecting virus itself, which depresses the host's immune response. Immunosuppression can be short-lived, but

it is always greater during the immune response phase.

2 Production of non-neutralizing antibodies or blocking antibodies: there are several viruses that persist in the host for life and induce the production of non-neutralizing antibodies. The antibodies react specifically with the viruses, but fail to render them non-infectious. These virus-antibody complexes can be demonstrated in the circulation. The non-neutralizing antibodies can block the activity of the neutralizing antibodies, thus favoring the persistence of the infection.

3 Insufficient viral antigen on the surface of the infected cell: if the antibody is to eliminate an infection by reacting with the viral antigen on the surface of the cell, fixing the complement and lysing the cell, then there must be a sufficient density of antigen for the process to take place. When there isn't enough antigen, the antigen-antibody interaction doesn't occur, so there is no lysis and the persistence of infection is favored. When the viral genome is integrated into the genome of the host cell and there is interaction between the two, the infected cells do not produce new infecting viruses and only the antigen formed induces a small formation of antibodies, which is not enough to completely fix and complete the process.

4 Direct cell-to-cell spread of the virus: antibodies can be ineffective and fail to cure the infection, because viruses spread directly from cell to cell, without passing through the extracellular fluid. An ineffective cellular immune response can also be responsible for persistent infections, both due to hyporeactivity of the cellular immune complex and the presence of blocked antigens, when antivirus antibodies combine with the viral antigen and block the antiviral activity of sensitized lymphocytes. On the other hand, there may be little viral antigen on the surface of the infected cell and the infected cells will escape the action of macrophages and lymphocytes, so the infected cell is not destroyed and the infection persists.

There are many viruses that persist in the host and are eliminated externally through saliva (Influenza, Herpes simplex, Cytomegalovirus), milk (Cytomegalovirus), blood (many viruses) and aerosols (common cold, Influenza) or urine, if the elimination of the viral antigen is restricted to this environment in the lumen of secretory channels, there is no way in which sensitized lymphocytes or circulating antibodies can reach them and exert their antiviral action. IgA antibodies can reach the virus or infected cells, but the reaction sequence with complement fixation is not activated and so the reaction is not completed. Although this mechanism is theoretically possible, it is not known how important it is in the persistence of viral infections. On the other hand, there may be a defect in the interferon response. Some viral infections do not lead to detectable interferon production, and are insensitive to interferon in some cells, but ineffective in inducing interferon production in others. In persistent infections, lymphocytes and macrophages are always infected, although it is difficult to conclude how these viruses behave in lymphocytes and macrophages. In a small number of viruses, macrophages appear to be the only infected cells. One possible significance of lymphocyte cells is that the host's immune response may be altered. Macrophages are the main cells in numerous viruses, and there are at least three ways in which infected macrophages may be responsible for persistent infection. Firstly, it has been shown that the interferon produced by infected macrophages may not have any protective value on other macrophages, although it does have normal activity on other cell types.

Secondly, although certain virus-antibody complexes are phagocytosed and degraded by macrophages, complexes by other viruses remain infectious for a long time after phagocytosis. Finally, macrophages and the reticulo-endothelial system are the main mechanisms for removing circulating viruses and virus-antibody complexes and there is a possibility that macrophages infected with certain viruses would be less effective in eliminating these viruses from the circulation, thus favoring the persistence of viremia. There are numerous data that can be invoked to explain persistent viral infections and it is hoped that new studies on the pathogenicity of viruses will be developed and guided by the mechanisms of virus persistence. Acute viral infections are being eliminated one by one through vaccination, and we will end up with only persistent infections. In some cases, satisfactory vaccination requires a rage of attenuated viruses which itself produces a persistent infection. In other persistent infections, when the host's defenses are overcome by viruses that persist intracellularly, in a non-infectious way but insinuated into the genome of the cell and the reproductive system, so that they are transmitted vertically, the problems of vaccination or cure are immense when the process of integration between the genomes of the virus and the host cell reaches a logical conclusion and makes it difficult to determine which is the virus, which is the host, the phenomenon is hardly recognized. On the other hand, if the host has benefited from the genetic contribution, the association can be classified as a symbiotic relationship and not as an infection. (ICN-virology 1976).

Chapter 3
3. Clinical manifestations

The usual clinical manifestations of influenza as an acute respiratory illness result from the infection and lysis of epithelial cells in the upper respiratory tract, trachea and bronchi. The virus enters the nasopharynx and passes through respiratory secretions. Viral neuraminidase hydrolyzes the mucoproteins. The incubation period is around two days. During the acute illness, the hair cells of the respiratory tract are necrotic. There is submucosal edema with infiltration of mononuclear cells and neutrophils. The most frequent symptoms of the disease are: dry cough, runny nose, headache, sneezing, anorexia, nasal discharge, fever, muscle pain, prostration and arthralgia; in more severe cases, nausea, vomiting, diarrhea and abdominal pain. The severity of the disease is associated with the viral concentration in nasopharyngeal secretions. Viremia is occasionally detected. The disease is self-limiting and lasts from three to seven days. From the third to the fifth day, the basal layer, which is affected, provides recovery of the epithelium. The severity and frequency of complications are variable and the virus can reach extensive areas of the lower respiratory tract, causing tracheobronchitis or pneumonia. The decrease in the number of alveolar macrophages and other cells that are infected by the virus can promote secondary bacterial infections, such as pneumococcus, streptococcus, Haemophylus influenzae and staphylococcus, commonly associated with fatal disease. Asthma, bronchitis and emphysema occur mainly in patients with cardiovascular defects.

Complications are rare, and pneumonia, myocarditis and pericarditis can occur. Neurological complications include peripheral encephalitis with polyneuritis, Guillan-Barre syndrome and Reye's syndrome which is a metabolic and neurological disease causing acute encephalopathy in children and adolescents, usually occurring between 2-16 years of age. Fatty degeneration of the liver is associated with this syndrome. The mortality rate is high (10-40%). The cause of Reye's syndrome is not known, but it is a recognized complication of influenza A and B virus infections. There is a possible correlation between the use of salicylates and the subsequent onset of Reye's syndrome. Although causal involvement has not yet been proven, it is advisable that children with flu-like symptoms should not be medicated with compounds containing acetylsalicylic acid for fever.

Many patients can tell you precisely when their first symptoms began. Initially, systemic symptoms predominate, which include fever, chills or generalized tremors, headache, myalgias, indisposition and anorexia. In the most severe cases, there is prostration. In general, myalgias or headaches are the most uncomfortable symptoms and their severity is related to the level of hyperthermia.

Patients commonly present with arthralgias. Although less common, ocular symptoms help establish the diagnosis and include photophobia, tearing, burning and pain when moving the eyes and redness. Respiratory symptoms, mainly a dry cough and nasal discharge, are usually also present at the beginning of the disease, but are outweighed by systemic symptoms. There may also be nasal obstruction, hoarseness and a non-exudative sore throat.

Fever is the most important physical abnormality. The temperature usually rises rapidly to a peak of 38 to 40°C and occasionally reaches 41°C within 12 hours of onset, accompanying the development of systemic symptoms. The fever is usually continuous but can be intermittent, especially if antipyretics are administered. On the second or third day of the illness, the temperature rise is lower than on the first day. As the fever decreases, the systemic symptoms subside. Typically, the fever lasts three days, but it can persist for 1-5 days or more. In the initial phase of the disease, the patient appears to be toxemic, the face is red and the skin is hot and clammy. Nasal discharge is common, but nasal obstruction is not frequent. The mucous membranes of the nose and throat are hyperemic, but there is no exudate. As the systemic symptoms and signs recede, respiratory complaints and abnormalities become more evident. Cough is the most common sign, it is bothersome and may be accompanied by substernal discomfort or burning. Nasal obstruction and secretions, pharyngeal pain and congestion are common. These symptoms and signs usually persist for 3 to 4 days after the fever recedes; however, coughing and malaise can persist for one, two or more weeks before complete recovery. Pharyngitis, croup and tracheobronchitis can also occur.

The diagnosis of influenza can often not be distinguished from infection with a number of other viruses and bacteria that cause headaches, muscle pain, fever and cough. Sometimes, other respiratory etiological agents produce an illness similar to influenza, such as streptococcal pharyngitis. In the summer months, enteroviruses produce a clinically indistinguishable picture and the acute manifestations of many other infections, such as dengue, can simulate influenza. On the other hand, in the context of an epidemic, influenza can be easily distinguished from other acute infections. When municipal, state or federal health authorities report an epidemic or infection by the Influenza A virus in a given community and a patient is seen with the acute onset of fever, headache, muscle pain and cough, it is highly likely that these symptoms are caused by an Influenza infection (Tarantino, 1990).

The clinical symptoms of influenza in children are similar to those in adults, although children have a higher fever and a higher incidence of gastrointestinal manifestations. Influenza A viruses are an important cause of croup (diphtheria) in infants under one year of age. In general, serious complications only occur in elderly and debilitated people, especially those with chronic cardiopulmonary diseases or other chronic diseases. Pregnancy appears to be a risk factor for lethal complications in some epidemics.

The lethal impact of a flu epidemic is reflected in the excessive number of deaths from pneumonia and cardiovascular and kidney diseases. Pneumonia as a complication of influenza virus infection can be of viral or secondary bacterial etiology, or a combination of the two. Increased mucus secretion helps to carry the agents to the lower respiratory tract. Influenza virus infection increases patients' susceptibility to bacterial superinfection. This is attributed to the loss of ciliary clearance, the dysfunction of phagocytic cells and the provision of an excellent bacterial culture medium by the alveolar exudate. The most common bacterial pathogens are Staphylococcus aureus, Streptococcus pneumoniae and Haemophylus influenzae. Pulmonary complications can be of 3 types: Primary Influenza Virus Pneumonia, Secondary Bacterial Pneumonia, Mixed Bacterial Viral Pneumonia. In addition, during an influenza outbreak, less distinct and milder pulmonary syndromes often occur which may represent viral tracheobronchitis, localized viral pneumonia or possibly mixed viral and bacterial infection.

Primary influenza pneumonia: this syndrome was well documented for the first time in the 1957-1958 pandemic. However, it is clear that many deaths in the 1918-1919 outbreak were due to this syndrome in healthy young adults. It occurred predominantly among people with cardiovascular disease, especially rheumatic heart disease with mitral stenosis.

Secondary bacterial pneumonia: bacterial superinfection is at first clinically distinguishable from primary viral pneumonia. Patients are often elderly or have pulmonary, cardiac, metabolic or other chronic diseases.

Mixed bacterial viral pneumonia: during an influenza outbreak, many cases are observed that do not fit clearly into any of the categories described above. The disease is not progressive, yet the fever pattern can be persistent. Patients may have a milder form of primary viral, secondary bacterial or mixed viral and bacterial infection. Many respond to antibiotics.

Milder forms of primary viral pneumonia involving only one lobe or segment of the lung have been described, which do not always lead to death. Such cases are more likely to be mistaken for pneumonia due to <u>Mycoplasma pneumoniae</u> than those produced by bacterial infection. Pneumonia can occur in children, but is less common than in adults. In addition, bronchitis and laryngotracheobronchitis are symptoms of this type of infection.

Chapter 4
4. Laboratory diagnosis

Infections with the common cold virus are diagnosed by clinical manifestations. Laboratory tests are not justified due to the wide variety of viruses and also because the disease is generally mild and self-limiting, without systemic spread. Diagnosis becomes important when the lower respiratory tract is involved, for example in influenza virus infections or in children infected with respiratory syncytial virus (RSV). The antigens of these viruses can be detected in the exfoliated cells present in nasopharyngeal aspirates, and a rise in the levels of virus-specific antibodies can provide a diagnosis (usually retrospective). Virus isolation is tedious and can be difficult, but when a pandemic is suspected of being caused by a new strain of influenza virus, it is usually carried out by central laboratories for public health purposes.

The laboratory diagnosis of influenza can therefore be obtained by isolating the virus or by serological tests. Definitive diagnosis depends on the isolation of the infecting virus or viral antigens in patient secretions or the detection of a serum antigen response. Influenza virus is readily isolated in nasopharyngeal, pharyngeal or laryngeal samples, sputum or tracheal secretion samples in the first two or three days of illness.

The success of virus diagnosis depends on the quality of the specimens collected and the conditions in which they are transported and processed in the laboratory. Specimens for the isolation of respiratory viruses in cell culture to directly obtain viral antigens and nucleic acids are usually found during the third day of the first symptoms and after clinical symptoms. The variety of specimens and ease of collection help in the diagnosis of infections by viruses found in the upper respiratory tract.

4.1 Types of samples that can be used:

a) Nasal swab: a dry swab should be inserted into the nostril parallel to the palate for a few seconds, making rotational movements, this should be done in both nostrils.

b)   Throat swab: both tonsils are cleaned vigorously with the swab.

c) Combined nasal and throat swab: the same procedure as in items a and b above should be carried out

d) Nasopharyngeal aspirate: the nasopharyngeal aspirate (NPS) is collected using a disposable urethral tube, which is attached to a catheter and a suction pump. The probe is inserted through the nostril into the nasopharynx and suction is applied, sucking the mucus into the collector. This procedure should be done in both nostrils. The collector is easily inserted by placing the patient in a supine position and immobilizing their head. After collection, the catheter is inserted into a vial containing 0.5mL of transport medium, medium 199 balanced salt solution with 5% gelatine, which is also aspirated into the collector to remove any secretion that may be stuck to the wall of the collector. After collection, the material obtained will be kept at 4°C until it is properly processed, so specimens should be transported on ice to the laboratory as soon as possible (Study et al, 1969).

e) Nasal lavage: patients are placed in a comfortable position with their head reclined backwards and told to keep their pharynx closed. They are asked to pronounce the letter K while the fluid is applied to the nostril. 1-1.5mL of lavage fluid is pipetted into each nostril using a protractor. The patient puts their head down so that the fluid can be poured into the beaker or petre dish. The process is repeated, alternating nostrils, until a total of 10-15ml of lavage has been obtained. Dilute approximately 3mL of lavage fluid 1:2 in transport medium.

f) Throat rinse: known as a mouthwash, the patient gargles 10mL of rinse fluid. The fluid is placed in a beaker or petre dish and diluted 1:2 with transport medium.

If clinically indicated, a diagnosis of respiratory virus of the lower respiratory tract can be made using the following methods: transtracheal aspirate; bronchoalveolar lavage and lung biopsy.

4.2 Influenza detection method

Standard laboratory methods for detecting influenza A virus are based on virus isolation and tissue culture (Frank and Couch, 1979) and (Monto and Massab, 1981). This method is also complemented by other practices for the diagnosis of infections caused by the Influenza A virus, such as the rapid detection of viral antibodies in nasal secretions by immunofluorescence or enzyme immunoassay (EIA) techniques. These tests are sensitive and required to identify viral antibodies in infected individuals. However, these serological methods are limited for the diagnosis of acute infection. Specimens used for direct detection of viral antibodies from infected cells should be stored on ice and processed within 1-2 hours. Specimens for isolated virus should be refrigerated immediately after collection and inoculated into susceptible cell culture as soon as possible.

Techniques for detecting, isolating and characterizing the virus or demonstrating the appearance of specific antibodies to viral antigens are described below:

4.3 Fixing Complement (Fc)

It is a simple test that can be used to diagnose a large number of viruses. This technique is based on the fact that an antigen-antibody complex fixes the complement added to the reaction by reacting dilutions of serum with a predetermined fixed amount of viral antigen and pre-titrated complement. After thirty minutes at 37°C, the hemolytic system (HS) is added to the reaction, consisting of a mixture of red blood cells (antigen) with rabbit anti-mouse red blood cell antibodies (hemolysin), where if in the first stage of the reaction the complement has been used (fixed) by the presence of the antigen-antibody complex (Ag-Ac) formed, the red blood cells sediment. If the complement has not been fixed, the RBCs will be lysed, indicating the absence of antibodies, i.e. the formation of the Ag-Ac complex (Downham, et al 1974).

Tab. 4.0 Complement fixation

STEPS	serum with ac	BLOOD WITHOUT Ac
1	serum(Ac)+Ag+C=(AcAgC)	serum(-)+Ag+C=(Ag+C)
2	AcAgC + SH= red blood cell sedimentation (+ reaction)	Ag+C+SH= lysis of red blood cells (reaction -)

Ac=antibody, Ag=antigen, C=complement (adapted from Oliveira, 1994)

4.4 Electron Microscopy (EM)

Although some authors have obtained good results using electron microscopy to diagnose Orthomyxoviruses (Doane, et al 1967) using nasopharyngeal secretions, few other authors have been able to reproduce these results (Evans and Olson, 1982). Its value is limited because it is very difficult to distinguish Orthomyxoviruses from Paramyxoviruses (Respiratory Syncytial Virus) in clinical material (Gardner, 1977). The EM technique has been widely used for the viral diagnosis of various clinical species, with excellent results. For the diagnosis of respiratory viral infections this technique is limited and for Influenza its use is restricted and used more for basic studies, not being recommended for viral diagnosis.(Word Health Organization-WHO 1981).

4.5 Hemagglutination Inhibition (HI)

A technique used to test for specific antibodies in serum, where the antibodies appear in the first week after infection, reach their maximum titers in 2 to 4 weeks, decrease slowly over a year, but persist at low titers for many years.

Hemagglutination refers to the hemagglutinin protein found on the virus envelope that has an affinity for agglutinating erythrocytes, hence its name. It is a traditional method for identifying influenza viruses. This method was originally described by Hirst in 1942 and later modified by Salk in 1944, who improved the test to be carried out on microplates. In general, the hemagglutinin antigen quantification standard is mixed with serial antiserum and red blood cells. Detailed below (see methodology).

Tab. 4.1 Inhibition of Hemagglutination

serum with ac	BLOOD WITHOUT Ac
serum (Ac) + virus (Ag) Ac + red blood cells (RBC sedimentation) result +	serum(-) + virus(Ag) = Ag + red blood cells (agglutination of red blood cells) result -

(Palmer, et al 1975

4.6 Enzyme-linked immunosorbent assay (EIA or ELISA)

The most commonly used enzyme-linked immunosorbent assay method for antigen detection uses a layer of capture antibodies on a solid phase to attach the antigen. The viral protein, if present, absorbs a second specific antibody which, in turn, is detected by an enzyme-linked antispecies globulin. The appropriate substrate is added for the enzymatic reaction. Enzymatic reaction products are obtained and the color developed can be qualified by a spectrophotometer or by simple observation by the operator comparing the color developed by the test samples with positive and negative standards. N-acetyl-cysteine or sonication is used as a mucolytic for nasopharyngeal secretions. The most commonly used enzyme is peroxidase.

Tab. 4.2 Immunoassay

DIRECT TECHNIQUE	INDIRECT TECHNIQUE
Antivirus Ac + specimen + enzyme-labeled antivirus Ac + enzyme substrate= Product Color	Using avidin-biotin (enhancer) Ac antivirus + specimen + biotin-labeled Ac + avidin + enzyme substrate= Product Color

4.7 Radioimmunoassay

The radioimmune assay was developed by Sarkinen et al, 1981 for the detection of influenza in nasopharyngeal secretions. The radioimmune assay proved to be as sensitive and specific as immunofluorescence, does not require laborious specimen processing and allows a large number of specimens to be tested each day. Several disadvantages are associated with the use of radioactive reagents: these include the relatively short half-life of radioactive reagents (Yalken, 1982) and (Shankkinen, et al, 1981). This technique detects the viral antigen or antibody in the clinical specimen. The Ag-Ac reaction marker is an enzyme and the reading is based on the color formed after the appropriate substrate is added to the enzyme used as the marker (alkaline phosphatase, B-galactosidase, peroxidase or urease). It is a technique used to detect the viral antigen or antibody in the clinical specimen, which can be feces, cerebrospinal fluid or nasopharyngeal aspirate.

Table 4.3 Radioimmunoassay

DIRECT METHOD	INDIRECT METHOD
Ac capture + specimen + Ac antivirus marked with I^{125} (read on cytometer)	Ac of guinea pig antivirus + specimen + Ac of rabbit antivirus + Anti IgG from$_T$ 125z, rabbit marked with I (read on cytometer)

4.8 In Situ Hybridization

The presence of viral nucleic acid can also be detected in secretory cells, swabs or biopsies using the hybridization technique. Such diagnostics have been used on a large scale to determine chronic viral diseases and can even replace simple virus isolation.

In the hybridization technique, the double-stranded nucleic acid chains of the sample and the chains of a standardized nucleic acid, the probe, are separated. The probe must have a minimum size of 12 nucleotides, which can represent the complete viral genome or part of it. After denaturation, the reassociation of these chains forms a hybrid, which can be visualized by marking the probe, which can be a radioactive isotope or an enzyme.

4.9 PCR - polymerase chain reaction

This technique makes it possible to selectively amplify DNA or RNA sequences, producing large quantities of DNA of a size and sequence defined from the target DNA you want to detect. The fragment of this technique is the enzymatic amplification of a DNA fragment flanked by two primers (oligonucleotides), which hybridize with the opposite strands of the target DNA. The orientation of the primers allows DNA synthesis by the polymerase to take place between them, duplicating the segment. The resulting products are also complementary and capable of joining nucleotides and each cycle doubles the amount of DNA synthesized in the previous cycle.

This technique is preferably used when working with small amounts of sample and should be coupled with DNA/RNA recognition techniques.

4.10 Immunofluorescence

The technique used for the detection of influenza has proven to be excellent for nasopharyngeal aspirates (Orstawk, et al.), in addition to the speed of diagnosis and the possibility of testing a large number of samples per day (Siqueira, et al.). The first use of immunofluorescence for viral diagnosis was through the examination of upper respiratory tract epithelial cell swabs during an influenza outbreak. This technique is widely used to detect antigens or antibodies to diagnose the presence of Influenza A and B, RSV, Parainfluenza types 1-4, Herpes 1 and 2, Varicella-zoster, Adenovirus, Rabies, Rotavirus, Measles and Eptein Barr virus (Oliveira, 1994). The success of the method lies in the quality of the specimen, the skill with which each slide is prepared, the specificity of the antiserum used and the experience of the microscopist (Mc, 1990).

This is the oldest and most comprehensive technique in the rapid diagnosis of viruses, where the marker of the Ac-Ag reaction is a fluorescent dye. It is based on the fact that the viral protein (Ag) is present in the specimen to be studied, it is captured by its homologous antibody (human immunoglobulin), forming the Ac-Ag complex and then the fluorescently-labeled anti-immunoglobulin is added, forming the fluorescent complex, which emits light when stimulated by a laser beam or intense xenon light. There are 2 types of tests covered by this technique: the direct method and the indirect method.

Tab. 4.4 Immunofluorescence

DIRECT TECHNIQUE	INDIRECT TECHNIQUE
Specimen(virus) + Ac labeled with dye Detection of fluorescence by UV light	Specimen(virus) + dye-labeled anti-Ac UV fluorescence detection

UV ultraviolet

4.11 Influenza virus cell culture

Growth, in the sense of cell proliferation, can be studied "in vitro", but the immense complexity of the metabolic, humoral and hormonal interrelationships that guarantee the homeostasis of biological systems must be taken into account. The availability of cell culture methods (primary, secondary and established lineages) that mimic the behavior of different tissues and organs, allows us to focus "in vitro" on cell proliferation, its control and the loss of control as a result of the phenomenon of neoplastic or malignant transformation that leads to the formation of a tumor (cancer).

Culturing cells "in vitro" is an art. Although cells can be influenced by the presence of each mutant, they become organized within the tissue. Cell culture makes it possible to observe cells in their entirety as well as to examine their component parts intrinsically.

Nowadays, cell culture is used in conjunction with cell biology and other molecular biology methods, breaking down the traditional boundary between the disciplines of biology and virology. Another is its methodology, where it is first necessary to determine the type of cell to be used in the study, since each organ and microorganism needs an appropriate type of cell for a particularized study of its interactions with other cells and the environment.

We have 3200 characterized cell lines, derived from another 75 species, including hybrids and plant cultures being studied and maintained in various national and international cell banks, including the American Type Culture Collection (Rockville, Maryland-USA).

a) Primary cells

Primary cells are freshly isolated and derived directly from the tissue of origin. The tissues of origin are mostly primary cell cultures from pathological species of animals and humans. Tissues from embryos of healthy animals are preferred to those of adults. The young animal guarantees better potential for successful cell culture. That's why there are so many different types of cells and different methods for cultivating them.

b) Aneuploid cells

Rat fibroblasts and cell cultures from a variety of human and animal tumors, otherwise not, sometimes become aneuploid (chromosome with multiple number not being their normal number), within the culture and in that case raising the continuum culture frequency. Continuous cell lines are generally aneuploid and have a complementary chromosome between the diploid and tetraploid values.

Normofibroblast cell lines are rare within cell culture lineages. A change in culture is common in continuous cell lines obtaining their transformation "in vitro" and sometimes spontaneously or chemically induced to another variety. The term "transformed" is well applied to the process of continuous cell lineage formation partly because the culture undergoes morphological and kinetic changes, but also because the formation of columnar cell lineages is sometimes accompanied by an increase in tumorigenicity (Freshney, 1983). The condition shown is arranged for the development of primary cells and continuous cell lineages; this probability is inherent in genetic variation, so it is not surprising to find genetic instability perpetuated in the continuous cell lineage.

c) Viral isolation in cell culture

Viral isolation in cell culture is the most sensitive method for diagnosing influenza virus(Johnston, et al 1990)(Siqueira, et al 1986), as it reveals the presence of the viral agent in the respiratory tract. Difficulties with the method are the need for viable viruses, with preserved infectivity in clinical material, as well as the need for cell culture, which requires laboratory equipment and experienced technicians.(Johnston, et al 1990)(W H O 1980).

The material obtained for the cell culture must be inoculated immediately into the culture, as the virus is highly labile and loses its infectivity rapidly, especially after freezing and thawing (McIntosh, et al 1990).(Walsh, et all 1989). Samples must be kept on ice or at 4°C from the moment they are collected until they are cultured. Isolation is done by observing the characteristic cytopathic effect of the influenza virus on the MDCK cell preferred for cultivation.

d) Cytopathic effect of influenza virus in cell culture

The general principles for collecting, developing and processing specimens, the choice of viral isolation system, the choice of test for diagnosis as well as the particularities of specimen culture have to be established before culture (Fields at al 1990, Hierholzer 1993, Liland and French 1988, Mandel at al 1990, Salmidt and Emmosns 1989).

The virus in culture is usually evidenced by the cytopathic effect in the course of single-layer cell infections. Cytopathic effect seen under a microscope at 40100X and observed in great detail at 200-400X. The exact nature of the cytopathic effect and the time required for the particular refinement of the cell type are indicative for a good study of the viral effect on the cell. Therefore, the observation of monolayers is important and their identification must be detailed.

The formation of the syncytium is accompanied by vacuolization during the cultivation of the influenza virus, and by granular degeneration due to the fact that the cells rarely separate (Castlls et al 1990, Franak et al 1979, Klend et al 1975, Meguro et al 1975).

Chapter 5
5.0 Epidemiology

The culture of influenza viruses directly from human species and serological detection indicate that there is a progressive improvement in the knowledge of how the influenza virus proceeds in the community. The laboratory plays a responsible role in the clinical diagnosis of influenza, with the probable existence of numerous other respiratory viruses capable of producing similar illnesses, as well as the identification of antigenic variations and mutations, Shift and Drift respectively, of the influenza virus on a global scale (Pereira, 1979). Similar variations are immediately recognized among the viruses circulating in a given region and their relationship with others is communicated, with Influenza prevailing as the most studied and examined.

The Influenza A virus has a wide range of hosts with natural interspecies transmission. It is distributed worldwide and due to its characteristic mechanism of genetic variation is responsible for the frequent occurrence of outbreaks, epidemics and pandemics (Dowdle et al 1975) (Pereira et al 1979).

In general, the incidence of influenza viruses in acute respiratory illnesses in children is lower than that of RSV, Painfluenza and Adenovirus (Glezen and Denny 1973) and (Winter et al 1996), but during periods of major outbreaks, they can be identified more frequently (Monto 1995). In Table 1.13 below, we highlight the clinical syndromes observed in respiratory virus infections.

Tab. 5.0 Most common clinical syndromes of respiratory virus infections

Agent	Incubation period	Most common syndromes
Influenza A, B and C	2-3	Flu
Parainfluenza 1,2,3 and 4	1-3	Runny nose, pharyngitis, conjunctivitis
RSV	3-5	Runny nose, croup, bronchitis
Reovirus	4-6	Runny nose, rhinitis
Adenovirus	3-5	Runny nose, pharyngitis, conjunctivitis

Coxsackievirus	6-12	Runny nose, pharyngitis and croup
Echovirus	6-12	Runny nose, pharyngitis and bronchitis
Rhinovirus	1-3	Runny nose, pharyngitis and croup
Coronavirus	4-6	Runny nose, pharyngitis and bronchitis
Herpesvirus 1 and 2	5-8	Pharyngitis

According to Veronesi, modified

The activity of the virus is monitored through a global surveillance program coordinated by the WHO and an international network of three reference laboratories in London, Atlanta and Melbourne. The program aims to monitor the circulation of strains and recommend the composition of vaccines (Hampsonm 1997).

Influenza virus activity worldwide during the period 1995-96 ranged from moderate to severe levels. The H3N2 subtype predominated in most countries in the Europe, some regions of the USA and China. It was isolated in Africa and was associated with an epidemic in Oceania during the months of April-June. During the period 1997-98, the Influenza A(H3N2) virus predominated in Canada and Europe during the months of January-February, in the USA from December 1997 to February 1998. It was isolated from outbreaks that occurred during the months of October-December/98 in Argentina, Chile and Fiji.

The H1N1 subtype caused epidemics in Japan and Africa, predominated in Canada, most of the USA and some European countries. It caused outbreaks in Santiago, Valparcúso and Brazil during the months of May-June and circulated at low levels in Europe, Taiwan, South Africa, Saudi Arabia causing outbreaks in Belarus, Russia and England.

In May 1997, the Influenza A(H5N1) virus, which usually only infects birds, was isolated for the first time in humans in a 3-year-old boy in Hong Kong. During the acute respiratory illness, the child presented multiple complications, passing into Reye's syndrome and died (VP DATSA: Influenza activity-wordwide, march-august/97, Cohen 1997).

According to Wesbter 1997, humanity will probably experience another human influenza pandemic at the end of 1999 and into 2000, and it will certainly originate in China. This biological event will be a consequence of the existence of virus reservoirs, genetically conserved in aquatic birds, and the mechanism of gene reassortment between human and avian viruses. Remember that there is also the possibility of an avian virus being introduced directly into the population without regrouping and causing an epidemic.

Influenza control in high-risk transmissions in the population depends on immunizations and the rapid identification of cases and control of these infections by institutions.

(Edwards and Gruber, 1991). The implementation of prophylaxis and therapeutic regimens is hampered by the inherent limitations of time-consuming methods of diagnosing influenza. As well as the implementation of prophylaxis and therapeutic regimens, this is hampered by the inherent limitations of time-consuming methods for diagnosing influenza.

Influenza is an international virus and does not respect geographical barriers or age, with minor and major epidemics having a considerable impact on the population. During the intervals between epidemics, it has been reported that the natural history of influenza epidemiology resurfaces in such a way that it is not possible to identify the virus and the cause of its emergence. The first essential step is a rapid complex verification of the problem and a follow-up to the discovery of the virus and its unestablished particularities.

Antigenic recombination offers a wide range of dissemination in the population in different age groups during the epidemic among the population. The consequences of an influenza epidemic and its spread through the population produce a number of international problems, requiring monitoring of these epidemic outbreaks and extensive research by scientific circles.

In order to reduce the severity of epidemics, attempts should be made to isolate the virus and identify the antigen it contains for possible monitoring. Identify a greater or lesser spread and produce a vaccine.

Throughout its history, epidemiology has proved to be an important and powerful tool for identifying and solving people's health problems. Part of this movement is the definition of needs, planning and evaluation of health services. All this effort to organize services and develop new technologies can be understood as a way for societies to understand their ailments and develop strategies to deal with them. Health services should be seen as a set of socially organized efforts to tackle the factors that generate diseases or, at least, to minimize their harm. However, throughout history, a fruitful debate has accumulated about the true results of these efforts in actually changing the health conditions of the population.

The health system is a complex that includes intense scientific development efforts, the industrial production process and, finally, the organization of its products into units (simple or complex) that provide services to society.

In the field of evaluation, the contributions of epidemiology, although relevant, still have many limitations. The capacity to produce and put into use new health care technologies (drugs, devices, procedures and organizational systems for health care) has grown exponentially. Alongside the potential for cure or prevention (not always confirmed) and the undesirable effects of these technologies are their high and rising costs, which are the cause of concern for all those with some responsibility for the health of individuals or populations. There is a mistaken view of social progress that it is the consequence of assimilating new technologies, leaving aside even their direct effects as generators of disease.

1.1 Antigenic variation

The virus has two antigenic variations, one with a major antigenic nature known as Antigenic Mutation (SHIFT) and another with a minor antigenic nature known as Antigenic Tendency (DRIFT). Both occur in Influenza A, B and the Drift effect has been observed in C and are related to the surface antigens Hemagglutinin and Neuraminidase.

1.2 Antigenic mutation (SHIFT)

This change causes the Hemagglutinin antigen to change completely and a new subtype to emerge. It is not necessary for this to also occur with Neuraminidase. It occurs at regular intervals. The first event was isolated in 1933 and a new one in 1957. These changes occur at intervals of 10-12 years. The human influenza pandemic was reported with the HswlNl virus at 19181919.

The appearance of new subtypes has certainly been associated with epidemiological behavior. Epidemics start in a particular place and spread to neighboring countries with a particular focal point. The appearance of a new subtype has been followed by a rapid disappearance of the old subtypes. However, events of this type of behavior are not mutable. For example, Influenza A from pigs showed a 75% capacity to infect humans over a period of time, causing a predominantly H3N2 epidemic in the United States. And most unexpectedly, it returned in 1977 as H1N1, but H3N2 didn't disappear. Both continued to circulate and were sometimes isolated in the same epidemic. The origin of new subtypes has been the subject of considerable speculation although the evidence is accumulating.

It has been suggested by Pereira,1969 that a host would be the reservoir of new viruses particularly in pure species in Asia where the last two subtypes have been seen. However, the actual reservoir remains to be identified. Another possibility is that a virus could occur through direct mutation of human species. This hypothesis would require multiple changes in the RNA coding for the Hemagglutinin antigen, as major differences have been found in the sequence of the Hemagglutinin amino acid chain of different subtypes. The possibility that these differences are the cause of these mutations has been small. There are more distant and experimentally supported possibilities that new subtypes are derived from the genetic interaction of existing human strains with influenza viruses from other mammalian and avian species. Many of these are infected with the Influenza A virus and the infection is mixed with more than one type of Influenza virus, leading to a high frequency of recombination of the viral particles. Some of these particles have hemagglutinin from one relative and another (Kilbourne, 1966) (Esaterday et al 1969) (Webster 1970). The ease with which many forms of recombination are naturally linked to the shape of the viral genome allows the rearrangement of RNA fragments, which can happen in vivo, as has been demonstrated experimentally under conditions similar to natural conditions (Webster 1973). some evidence that one of the many A/Hong Kong/68(H3N2) subtypes may have arisen as a result of genetic recombination between the earlier predominant Asian H3N2 viruses and a non-human chain reported for A/duck/Ukraier/62(Hav7

Neq2/A/equine/Miame 63(Heq2Neq2) viruses(Laver and Webster 1973).

1.3 Antigenic Trend (DRIFT)

During the prevalence of the influenza virus of the subtype of lower antigenic nature, changes found in surface antigens are irregular with longer intervals of change (Shift). The antigen drift mechanism appears to involve immunological selection of a mutant virus with an alteration in a given antigen. Such chains are produced in the laboratory by culturing the virus in the presence of limited dilutions of specific antibodies (Laver and Webster, 1968). Mutants have also been found with chains in the amino acid polypeptide sequence of Hemagglutinin and similar chains have been demonstrated during a natural drift antigen. This Drift antigen has been observed in each of the last two decades, variants of the H2N2 and H3N2 subtypes began to appear after the third major epidemic when a considerable percentage of the population would have been infected in the H2N2 decade where its variants were detected.

In the decade of H3N2 a similar type was observed in 1978 and since then five variants have shown a potential for special epidemics, during this time other variants that showed a similar degree of antigenic differences have been isolated, but for unknown reasons one virus has failed to spread and simply disappeared.

A future contribution that the laboratory can make to the survival of influenza lies in the study of the immune potential of the population where the information can be collected by this means and invalidated from many aspects when there is no continuation of this collection. Firstly, the regular collection of serum samples before and after influenza epidemics can be used as a measure to study the spread of the current subtype or variant, which can be particularly useful where other epidemiological data is not feasible.

The use of serology to highlight the value of cross-reacting antibodies when variants begin to appear according to the prevalence of influenza can be measured indirectly by collecting and analyzing information on mortality and morbidity. There is a lot of information from many of the developed countries with serological evidence of the presence of the Influenza virus circulating in the population. In many countries of the world, however, there is no statistical data or influenza has not passed through the population except when a serious epidemic occurs. Some of these countries are in the southern hemisphere, where influenza typically occurs in the winter, which corresponds to the summer months in the northern hemisphere, a time normally free for influenza. This alternation of predominance is important for research into epidemic events in the northern hemisphere.

Tab 5.1 Hemagglutinin(H) and Neuraminidase(N) antigenic subtypes of human influenza A virus

prevalence period	subtype of H	reference chain	subtype of N	reference chain
1918-1919	Hsw1	Hsw1N1	N!	Hsw1N1

1962	H3	A/duck/Ukrai er/62 (Hav7Neq2/A/equine/ Miame63Heq2Neq2 A/ H3N2)	N2	A/duck/Ukrai er/62 (Hav7Neq2/A/equin e/ Miame63Heq2Neq2 A/ H3N2)
1968	H3	A/Hong Kong/68 (H3N2)	N2	A/Hong Kong/68 (H3N2)
1977	H1	H1N1	N1	H1N1
1978	H3	A/H3N2	N2	A/H3N2

Chapter 6
6. Morbidity

Influenza does not produce its own clinical characteristics and its recognition even during the outbreak of an epidemic is difficult without laboratory examination. In epidemics of acute respiratory diseases in military camps, for example, it has been difficult to differentiate an infection with the influenza virus from one with another respiratory virus (McDonald et al 1962). Data for influenza morbidity is based directly on data for acute respiratory diseases.

Many years of experience have shown that there is a correlation between the increase in morbidity data and the presence of influenza in the community. Acute respiratory infections are of major significance. Suffice to say that, of all the diseases that attack man, they are the most frequent and it is well established that 90% of respiratory tract infections are of viral origin (Paterson, 1975).

Acute respiratory infections are of significant importance. Suffice to say that, of all the acute diseases that attack man, they are the most frequent and it is well established that 90% of respiratory tract infections are of viral origin (Paterson 1975). According to estimates by Jordan and Cols (Jordan et all 1956), there are twenty billion cases of acute respiratory diseases in the world every year and each individual, on average, has approximately six respiratory infections, in either benign or severe forms. In the USA, there are 500 million cases of respiratory infections a year, 10% of the population is affected, meaning that 20 million people have the most serious economic implications, even representing a serious social disruption factor through absences from work, absences from school and repercussions on industry. Hence the existence of studies on the econometrics of health, or perhaps it would be better to say the econometrics of illness, seeking to assess the damaging effects in economic and financial terms of these constant illnesses. To give an accurate picture of the economic implications of such infections, it is enough to mention that during the 1957 flu pandemic in England, 12 million people were affected, a quarter of the country; there were 16,000 deaths and the loss to British industries was in the order of one hundred million pounds sterling (Góes, 1954). There are also indicators, in the form of financial figures, in US data, which tell us of equivalent effects (Lepine, 1964). There are relatively frequent cases in our country with a clinical diagnosis of viral etiology. However, this is mere speculation, as the viral origin is not usually confirmed by isolation or serology.

As seen above, the vast majority of respiratory infections are caused by viruses, from which it can be inferred that these are the main agents of human morbidity, since respiratory infections are in the foreground and among them those caused by viruses stand out.

The viruses that cause respiratory infections include almost all the groups of viruses known today. The modern classification of viruses pathogenic to humans comprises 15 major groups (Andrewes et all 1979), namely DNA viruses, adenoviruses, which cause major respiratory infections, herpesviruses which, in addition to various syndromes, can cause pharyngitis and other respiratory infections; the RNA picorvaviruses, small viruses, which include the Rhinoviruses, which cause colds, the Coxsackieviruses, the Echoviruses, both of which belong to the Enteroviruses, and the polioviruses, which sometimes cause symptoms, which the Americans call influenza-like. We also have the Orthomixoviruses, which include the Influenza viruses, the Paramixoviruses, which include the Parainfluenza viruses and the Respiratory Syncytial Viruses, and the Coronaviruses. Our diagram shows the "main" groups of viruses, but we mustn't forget the "secondary" groups or subgroups, as it turns out that half of the known viruses are involved in the etiology of respiratory infections.

The main respiratory viruses and their corresponding clinical entities are listed: Influenza A, B and C viruses cause influenza in its typical form, but they also cause complications such as pneumonia and meningoencephalitis. Parainfluenza, on the other hand, occurs in children and causes coryza, pharyngitis, otitis, croup and bronchitis; respiratory syncytial viruses, which cause mild symptoms in adults, but in children cause more severe forms, such as croup, bronchitis, bronchiolitis and pneumonia, in which case they often occur in young children, between 3 and 5 months old, due to a worsening of the infection due to antibodies of maternal origin transferred via the placenta. As a result of this phenomenon, there is a combination of virus-antibody and complement within lung cells parasitized by the virus. This complex causes hyperplasia and cellular inflammation with serious consequences (Christie, 1969)(Evans, 1976), which can often lead to the patient's death.

The Influenza virus, of all the agents mentioned, is perhaps the most attractive because it has the widest incidence, occurring in epidemic manifestations, sometimes alternating, for longer periods, with pandemics, assuming great virulence in these epidemic and pandemic incursions. This virus is characterized by extreme morbidity, a communicability that may not exist in any other infecting agent and, in addition to the less serious manifestations of conventional clinical pictures, it is responsible for pneumonic syndromes and possibly, rarely, also manifestations of aseptic meningitis, meningoencephalitis (Kilbournr, 1975). Influenza is a problem, with episodes that recur monotonously being lost to history. This is because influenza viruses undergo permanent immunological changes, which are of two types: the so-called immunological drift, which are secondary changes, and the immunological shift, which are profound or major changes, in which the predecessor sample, from which a new mutant descended, has no antigenic relationship whatsoever, and does not present cross-immunity of any kind with the new mutant that has appeared(Dolpheide, et all 1980)(Gething, 1979)(Ward, at al 1980).

Possibly they cross in nature with certain group A viruses from horses, pigs and birds, and these wild animal samples compete for new antigens from circulating human viruses, giving rise to mutants with different characteristics. This phenomenon is now very well known and information about it is based on antigenic analysis. In this way, once we have determined the antibodies of individuals of a certain age, of old people, or of sera stored in serum banks, we can find out what happened many years ago and the type of virus that circulated at that time. It was through these studies of sera - retrospective epidemiology - that we were able to determine the characteristics of the sample that circulated from 1874 to 1889, which is Equi2/Miami 63, a relative of an equine influenza virus. The sample from Asia 57 (Góes, 1959), A2 was that of the 1901 epidemic, a conclusion drawn from the examination of serum from individuals from that period and the existing antibodies to Asia 57. That is, before the discovery of the viruses, it was possible by inference to gather data from these retrospective studies. In 1902-1917, a sample similar to the one known as A2/Hong-Kong/68 circulated, and so on. From 1933, when AO/PR 8/34 was isolated (Smith, 1933), by comparison with isolated samples, it became possible to carry out antigenic analysis and retrospective research with the sera of older individuals who had suffered infections at a time when the influenza virus was not yet known. With these procedures, it was possible to reach a conclusion about the antigenic make-up of influenza agents that had occurred previously.

As we have seen, acute infections represent the most frequent morbid states in human pathology and morbidity statistics indicate that these diseases have the highest rates. However, the extremely high frequency of respiratory infections can be explained by isolation and serological tests.

There are some interesting aspects to immune phenomena. As we have seen, there are some paradoxical facts about immunity in respiratory infections. An individual, for example, who has antibodies against respiratory syncytial viruses, especially children, more often develops severe bronchitis and pneumonia than those who don't have antibodies, because these antibodies, combined with the viruses, form immune complexes that fix complexes, resulting in a damaging process at the alveolar level.

There are also facts that have not been properly clarified, such as the fact that humoral immunity depends on IgG, which in some cases has no significance in antiviral protection. So much so that even in the case of respiratory syncytial virus, newborns with high titers of the mother's original antibodies are extraordinarily susceptible to RSV, meaning that blood IgG has no protective action.

Epidemiological factors explain the high frequency of respiratory infections. Airborne infections are the most widespread. An outbreak of influenza that appears in the East, Hong Kong for example, three to four months later has already invaded the world, and it is not known how or by what mechanism. However, oral/fecal infections, such as cholera, can be confined to a single continent and can therefore be contained by vaccination and sanitation measures. But there are no effective resources for suppressing airborne infections. That's why this situation can only be described as dramatic, given that acute respiratory infections dominate all human pathology.

6.1 Mortality

Influenza mortality data can be expressed on a weekly basis by the number of deaths due to influenza, or due to bronchitis, pneumonia or as respiratory diseases (Languir and Housworth, 1969; Assad and Reid, 1971) but, although mortality can be measured as a more objective index than morbidity, the attribution of each disease caused by influenza is similarly based on clinical judgment and is prone to error, which is why groundwork is needed in the scientific field for follow-up in cases of influenza outbreaks. However, the concept of total excess mortality as a reflection of influenza activity, as originally suggested by Far in 1985, has been confirmed as a disease with a higher global mortality rate (Assad et al 1973).

The index can be used as a retrospective measure of the impact of influenza in any given winter by estimating the total number of excess deaths even when no influenza virus has been isolated despite routine virological research, or it can be applied to events occurring by calculating an epidemic which follows the recognition of a significant increase in the number of deaths extracted from weekly mortality analyses of deaths in previous years. This method can be applied in those countries where deaths are certainly notified without delay, but this is still a long way from becoming common practice in many areas of the developing world. If it were possible to stabilize the method, the simplest way to do so would be to compare the prevalence of influenza in different parts of the world. Mechanisms that generate the spontaneous cure of viruses in the selection of species in nature (appendix 01)

Chapter 7
7. Serological surveillance

The use of surveillance, especially in the last four decades, has expanded significantly, especially since the eradication of smallpox in the 1960s, when its use spread to all continents, consolidating it as an important tool of epidemiology in health services.

Because this experience has taken place in numerous countries with varying degrees of economic development and with different types of health system structure and organization, we have seen the incorporation of surveillance in the different regions of the world, with particular characteristics, sometimes restricted to the role of an information system for speeding up control actions, often being confused with the control of communicable diseases themselves, sometimes encompassing research and being confused with epidemiology itself.

Its methodology can be summarized as the activity of continuous monitoring and regular analysis of the behavior of specific adverse health events in populations and the elaboration, based on scientific knowledge, of the technical bases that support the strategies adopted by the programs to control these events.

To make the terminology used clearer, we understand public health instruments to be the resources used to achieve the goal of providing comprehensive public health care: programmatic action, planning, health education, health inspection, surveillance, monitoring, public health research and human resources training.

Most influenza programs have emphasized the rapid identification of laboratory-isolated influenza viruses around the world. The laboratory aspects of influenza epidemiology are certainly the most widespread and uniformly covered in the world. Since they depend on many individuals they have been well treated and tested with WHO laboratory facilities and agents. The functions of national laboratories are: to make and maintain connections with medical clinics and health centers, schools, private and public industries to obtain clinical specimens isolated from patients with acute respiratory disease.

Surveillance has two internationally recognized meanings in public health: one that we could call classic, first applied at the end of the last century, is linked to the concepts of isolation and forty. These concepts emerged at the end of the Middle Ages and were consolidated in the 17th and 18th centuries with the strengthening of commerce and the proliferation of urban centers.

Isolation and quarantine determined the separation of individuals from their usual contacts, taking on a compulsory character, typical of the medical police, aimed at defending healthy people, separating them from the sick or those who could potentially become sick. This set of restrictive, police and punitive measures created serious difficulties for trade between countries.

From the 1950s onwards, we saw the introduction of a new concept of surveillance applied to public health, this time in the sense of systematic monitoring of adverse health events in the community, with the aim of improving control measures. The methodology applied by surveillance, in this new concept, includes the systematic collection of relevant data relating to specific adverse health events and their continuous evaluation and dissemination to all those who need to know about them.

Since 1989, the term epidemiological surveillance has been replaced by public health surveillance, which has become internationally established and has been used in all publications on the subject since the early 1990s, including a recent document from the Pan American Health Organization. *It should be emphasized that this change of name did not imply the adoption of a new approach or changes in the conceptual or operational aspects of surveillance.*

7.1 The operational aspects of surveillance

In order not to limit ourselves to conceptual aspects in the characterization of surveillance as a public health tool, we will now briefly present some operational aspects that characterize surveillance as one of the applications of epidemiology in health services.

7.2 Objectives of surveillance

The main objectives of surveillance include:

1- Identify trends, groups and risk factors in order to develop strategies to control specific adverse health events.
2- Describe the pattern of occurrences of diseases of public health importance.
3- Detecting epidemics.
4- Documenting the spread of diseases.
5- To estimate the magnitude of morbidity and mortality caused by certain diseases.
6- To recommend, on an objective and scientific basis, the necessary measures to prevent and control the occurrence of specific health problems.
7- Evaluate the impact of intervention measures.
8- Evaluate the suitability of tactics and strategies for applying intervention measures, not only in terms of their technical foundations but also those relating to the actual implementation of these interventions.

The objective of surveillance is not merely the collection and analysis of information, but also the responsibility to develop, based on *rigorously updated scientific knowledge, the technical bases* that will provide support for health services with the concern for continuous improvement, as well as speeding up the identification of problems in order to provide timely intervention for their control.

Chapter 8
8. Objective

^   To determine the immunogenic capacity of the new influenza vaccine containing two strains of type A and one of type B to be administered to the study population, which consists of patients aged 60 and over, in April 2000, in Maceió-AL, as recommended by COLAB (attached);

^   Monitor the circulation of other influenza virus strains;

^   Determine the incidence of the vaccine, i.e. whether or not it prevented an influenza outbreak;

^   The data obtained was notified to the Epidemiological Surveillance Service of the Alagoas State Health Department.

Chapter 9
9. Material and Method

9.1 Material

Calcium-saline solution

Weigh and dissolve in 1L of distilled water

Calcium chloride bi-hydrate ($CaCl_22H_2O$)	1 gram
Anhydrous Sodium Chloride (NaCl)	9 grams
Boric acid (H_3Bo_3)	1.2 grams
Anhydrous sodium borate ($Na_2B_4O_7$)	52 grams

After dissolving the salts, filter or autoclave to sterilize; the Ph of the solution should be adjusted to 5.6.

9.1.2 Sodium Citrate Solution

Weigh and dissolve 1.6 grams of Sodium Citrate bi-hydrate ($Na_3C_6H_5O_7/2H_2O$) in 100mL of PBS 7.2. Filter or autoclave to sterilize.

Note: Calcium-saline and sodium citrate solutions are stored at room temperature and 4°C respectively.

9.1.3 Alsever's solution

Note: Autoclave 8-9 pounds for 10 minutes and store at 4°C.

Destrose(anhydrous) C6H12O6	20.5grams
Sodium Citrate Na3C6H5O7 2H2O	8 grams
Citric acid H8C6O7H2O	0.55 grams
Sodium chloride NaCl	4.2 grams
Distilled or demineralized H2O	1000mL

Through laboratory diagnosis of influenza obtained by serological tests.

9.2 Sample used

Vacuum blood collection is widely used to obtain blood samples without the use of a syringe. In addition to the ease and speed of collection, the system has a positive impact on the quality and preservation of the samples.

Venous blood is used for most laboratory tests. It is collected using vacuum tubes fitted with sterile needles, with or without anticoagulants. 25x8 or 25x9 needles were used to collect the blood. Venous puncture is usually performed in the cubital vein, in the fold of the forearm, in adults, which is preferred for this type of procedure.

The steps for collecting blood from the cubital vein are described in Appendix 02.

9.3 Method used

The technique used to test for specific antibodies in serum was Hemagglutination Inhibition (HI). Palmer et al 1975 where the antibodies appear in the 1st week after infection, reach their maximum titers in 2 to 4 weeks, decrease slowly over a year, but persist in low titers for many years. Hemagglutination refers to a protein, Hemagglutinin, found on the virus envelope that has an affinity for agglutinating erythrocytes, hence its name. It is a traditional method for identifying the influenza virus. This test was originally described by Hirst in 1942 and later modified by Salk in 1944, who improved the test to be carried out on microplates. In general, the quantified Hemagglutinin antigen standard is mixed with serially diluted antiserum and red blood cells from chicks. However, some considerations about this technique are necessary:

9.3.1 *Obtaining chick red blood cells*

1. Use one-day-old, unfed chicks.
2. Anesthetize the chicks with chloroform.
3. By cardiac puncture, bleed the chick with a 2mL syringe (in the ratio of 1.2mL of

alsever to 0.8mL of blood).

4. After collection, add the contents of the syringe to a vial containing 15mL of alsever (the contents of 5 syringes can be added to each vial).

5. Store the vial with the red blood cells at 4°C until use.

9.3.2 RBC washing

1. Centrifuge the RBCs in a graduated conical tube for 10 minutes at 1000 rpm.

2. Discard the supernatant, resuspend the sediment in PBS and centrifuge again (repeat this procedure once more).

3. Dilute the sediment 1:2 and store at 4°C until use.

9.4 Hemagglutination Inhibition Considerations

a) The HI test is used to determine specific antibodies against influenza viruses using reference antigens.

b) The reference antigens to be used must be from the strains of the Influenza A and B virus that are circulating at the moment (they are usually contained in the latest kit sent out by the C.D.C. Center for Disease Control).

c) Patient sera as well as homologous sera must be pre-treated with RDE (enzyme receptor destroyer).

d) Every HI test must be preceded by titrations (by hemagglutination) with the viruses to be used. The aim of the first titration is to find out the titer of the virus and to determine the dilution containing the 8 hemagglutinating units needed to carry out the HI test. The dilution containing the 8 hemagglutinating units is obtained by dividing the "end point" dilution (the highest dilution of the virus with complete hemagglutination), for example, if the virus titer was determined to be 640, after dividing we will have the 1:80 dilution, which is the dilution containing the 8 hemagglutinating units. Once the dilution for use has been determined, a sufficient quantity of this dilution must be made to carry out the entire test. The second titration, carried out just before the HI test, aims to confirm the concentration of the virus in 8 units. This titration should be carried out using the virus already diluted to the concentration to be used. Finally, the third and last titration will also be carried out with the virus (at the concentration of use) after the end of the HI test to check that there has been no drop in titer during its use.

e) Every HI test must be accompanied by a homologous system (each virus used must be tested with its corresponding antiserum; for example, virus A/Brazil/11/78 with its antiserum A/Brazil/11/78).

9.5 Serum and reference treatment for elimination of non-specific inhibitors by RDE

1 - Add 0.1mL of the serum to be treated to a 12 x 75mm tube and label it.

2 - Add 0.4mL of RDE (diluted 1:20 in calcium saline) to the tube with serum.

3 - Shake and cork the tube, taking care to seal the stopper to the tube with adhesive tape.

4 - Incubate in a water bath at 31°C overnight.

5 - Add 0.3 mL of 1.5% sodium citrate solution.

6 - Repeat the procedure from item 3.

7 - Incubate in a water bath at 56°C for 30 minutes.

8 - Add 0.1 mL of chicken or chick liver flukes (diluted to 50% in PBS, Phosphate Buffer 7.2).

9 - Incubate for 30 minutes in the refrigerator (4°C).

10 - Centrifuge at 1000 rpm for 10 minutes in a refrigerated centrifuge.

11 - Transfer the supernatant to another tube and discard the sediment.

12 - Add 0.2 mL of PBS 7.2.

Watch:

The serum after treatment can be stored at -20°C.

The serum at the end of the treatment will result in a final dilution of 1:10.

Diluted RDE (1:20) can be stored at -20°C.

9.6 Antigen titration

1 - Use U-bottom microplates for testing.

2 - Mark the microplates using a row of holes for each antigen, leaving the last hole for the RBC control.

3 - Add 1 drop of diluent (PBS pH 7.2) to all the holes in the plate except the first hole.

4 - Add 2 drops of the virus to the first hole and 1 drop to the second and third holes.

5 - Dilute with loops or a micropipette from the third to the tenth hole of the microplate.

6 - Add 1 drop of PBS 7.2 from the third hole, including the RBC control.

7 - Add 2 drops of chicken or chick red blood cells (diluted 1:200 in PBS 7.2) to all the holes.

8 - Incubate for up to 30 minutes at room temperature, observing and determining the highest dilution at which the virus causes complete hemagglutination of the red blood cells.

POSITIVE RESULT/HEMAGGLUTINATION.
NEGATIVE RESULT/HEMOSEDIMENTATION

9 .7 Running the test

1 - Treat reference antisera and sera from patients with RDE.

2 - On the day of the test, wash the chicken, chick or goose RBCs in PBS 7.2 and then dilute them 1:200 (make enough for the whole test).

3 - Perform the first titration of the virus, find the end point and determine the dilution that contains the 8 hamagglutinating units.

4 - Dilute the virus in the dilution determined previously (make enough for the whole test) and then carry out the second titration.

5 - Once the stability of the virus in the dilution has been confirmed, mark the microplates with the records of the patients' sera and the homologous serum.

6 - Start the test. Drip 1 drop of PBS into all the holes except the first.

7 - Add a drop of RDE-treated serum to the first, second and control orifices.

8 - Dilute with loops or a micropipette from the second hole.

9 - Add 1 drop of the virus dilution containing the 8 hamagglutinating units into all the holes except the control hole.

10 - Incubate the microplate for 30 minutes at room temperature.

11 - Add 2 drops of witch hazel (1:200 in PBS) to all the holes.

12 - Incubate for 30 minutes at room temperature and take the reading

Tab 9.0 HA test scheme for influenza

Diluições						
∅	O		2 gts de Ag			2 gts de gobs
1:2	O	1 gta de PBS	1 gta de Ag 1 gta de Ag	Diluir	1 gta de PBS	
1:4	O					
1:8	O					
1:16	O					
1:32	O					
1:64	O					
1:128	O					
1:256	O					
1:512	O					
C	O					

Chapter 10
10. General characteristics of the city of Maceió

10.1 History

The Tupi Indians, amazed by the natural phenomena that occurred here, called it "Maçayô" or "Maçai-ok", which means "What covers the swamp". And so the name Maceió came about.

The town originated as an old sugar mill around the 18th century. Its development began with the arrival of ships carrying wood from the Jaraguá inlet. With the emergence of the mills. Maceió began to export sugar, then tobacco, coconut, leather and some spices.

Prosperity led to the town becoming a village on December 5, 1815. Thanks to its growth, on December 9, 1939, Maceió was already the capital of the province of Alagoas.

Today, the grandeur of a bygone era is still present in the Jaraguá neighborhood, with its port and mansion, whose architecture makes us travel back in time.

The city of Maceió is located at 05 meters above sea level, its altitude is 16 meters and it covers an area of $508km^2$ with a population of 723,230 inhabitants. The climate is temperate and the rainiest periods are from June to December and the least rainy from January to May. To the east of the state of Alagoas, the climate is tropical, semi-warm and semi-humid.

Main access roads: BR 101, BR 104 and BR 115.

The average temperature is 24°C and the relative humidity is 85%.

Information Guide to the Municipalities of Alagoas 1999.

Appendix 1

The mechanisms that generate spontaneous healing: The influence of viruses on species selection in nature

According to the WHO, 2/3 of the infectious diseases that affect humans and animals are viruses, so the notion that man is surrounded by viruses is an artistic vision, but it doesn't escape reality. Another well-established point is that there is no specific treatment for viruses. For many of them there are forms of prevention, and for the vast majority there are control measures. So what we have is that people become infected in various ways, whether by exercising polite habits, such as wishing their neighbor good health, or in an uncontrollable outburst of affection, or in some other less pleasant way such as an injection, transfusion or other. Some of these people die as a result of the virus, but most get well and will never develop the same virus again. Of course, someone can say that they've had the flu "n" times, but it certainly wasn't the same kind. So it's quite relevant to talk about a spontaneous cure. So what are the mechanisms that allow this cure to happen? Some people attribute the cure to antibodies, but if antibodies really had this function, we wouldn't be dealing with the so-called "plague of the century", because we all know that the laboratory diagnosis to identify an individual as HIV+ consists of testing for antibodies which, although present, do not prevent the individual from evolving into AIDS.

A careful analysis shows that the antibodies are really important in a second episode, but the first time there is contact with the virus, the antibodies appear with the decline of the virus. So what makes viruses self-limiting?

Interferons (IFNs) are a family of cytokines that were first detected and named by Isaac and Lindenmann in 1957 because they interfered with the production of viruses by chicken embryo cells. Since then, numerous groups have dedicated themselves to the study of these elements and, currently, we have established the IFNs a, P, Y AND even o and co_, the latter having a similarity to a. The main properties of IFNs are their structure, the cells that produce them and others.

In order to synthesize viruses from the genetic information they have captured, non-lymphoid cells go into intense anabolism in order to perform their new function. At the same time as synthesizing viruses, they also synthesize IFN P. The greater the number of cells in the stage of competence for producing viruses, the more widespread the virus will be: For example: $G1^{\wedge}S^{\wedge}G2^{\wedge}M$ If the enzymes necessary for viral biosynthesis occur in G1, the greater the number of cells in G1 in a given organ or tissue, the greater the damage to that organ.

While the tissue damage is being produced and eliminated by the body's cleansing elements, the lymphocytes have activation factors at their disposal and, in order to proliferate, they need food (glucose and insulin receptors). As they are dependent on or controlled by the level of plasma corticoids, in order to proliferate they begin to use ACTH (corticotropin) receptors and their medullary activity is diminished renewal of non-lymphoid cells compromised by available nutrients and an increase in the number of lymphocytes. By ending the Ags population without nutrients and decreasing uptake, the cells enter apoptosis and, at this stage of programmed death, their waste products serve as nutrients for the other cell populations.

The level of ACHT returns to normal, with the consequent production of corticoids, leading to a decrease in lymphocytes and the healing state is characterized by the recording of contact (Ac) and memory cells.

A man with interferosis is a person with a typical acute viral illness. But anyone who has seen a person injected with cloned IFN cannot distinguish the symptoms, so there is a certain sense in calling this condition interferosis: IFN at high levels cell renewal is not occurring at a normal level, the recomposition of cells in the blood vessels, joints, skin and those of the myeloid series, is in "slow motion", with a decrease in platelets, granulocytes, monocytes and consequently M0. In this state of malaise, it is clear that reflexes and senses are very heightened. Just like humans, animals are also subject to viruses and, consequently, interferosis. Let's imagine a flock of birds looking for a better climate in which to mate and breed. If some of these birds become infected with viruses and have the characteristic of having many cells competent to make viruses, what will happen? They become inattentive and thus easy prey for their predators. Only the fittest make it to the end, i.e. those with the least competence, with their genetic characteristics.

You don't see sick animals in the wild, unlike humans and their pets. Another example is what happens to whales, leading to isolated or mass strandings.

Humans have no predators, so they have their viruses and survive by breeding, and some even exaggerate. According to the forecasts, overpopulation in the third millennium will be a fact, which in some places is already being felt.
The limiting factor in this case will be the lack of food and
water, as is the case
with Australian rabbits. Brazil clearly has an
advantage,

Appendix 2

The established steps for collecting blood from the cubital vein are:

1- first check that the vacuum harvesting system is suitable;

2- place the tourniquet around the patient's arm, above the elbow bend. Check the pulse to ensure that arterial circulation is not interrupted (especially in the case of the population studied);

3- ask the patient to open and close their hand several times to increase venous circulation (this cannot be done when collecting various hematological tests);

4- By inspection and palpation, determine the vein to be punctured, which should be large and firm. It is very important to learn to feel the presence of the vein by palpation. Often, in children or obese patients, good sensitivity to palpation locates veins that are not visible. Care should be taken with frequently bleeding patients and, in our case, the elderly, as the veins chosen may be totally or partially obstructed, from which no material can be collected.

5- Disinfect the skin over the selected vein with 70% alcohol and allow to dry before puncturing.

6- Do not touch the area to be punctured and do not let the patient bend their arm.

7- Ask the patient to keep their hand closed.

8- Take the syringe and place your finger on the needle mandrel to guide it during insertion into the vein.

9- Stretch the skin of the elbow crease with the index finger of the other hand, about 5 cm below the site of the puncture, but without touching it.

10-Insert the needle into the skin next to the vein to be punctured, parallel to it, and slowly penetrate inside. A slight loosening of the resistance to the needle's penetration indicates that it has been inserted into the vein. However, this loosening is not always perceived

11-The blood will flow into the syringe spontaneously. If the embolus is very stuck, as is the case with most plastic syringes, or if the blood pressure in the punctured vein is low, we should pull lightly on the embolus to check that the needle is in the vein and then withdraw the necessary blood.

Be careful not to transfix the vein. If the vein is transfixed or the needle is pulled out of it, we don't get the required volume of blood, blood leaks into the tissues and a hematoma forms. The blood from the hematoma cannot be used because it will be hemolyzed. If there is a hematoma, ask the patient to open their hand, immediately remove the tourniquet, remove the needle, place a dry cotton pad on the puncture site, make the patient bend their arm and look for another vein, if possible on the other arm.

12-In the normal collection, once the required amount of blood has been drawn, loosen the tourniquet, remove the needle and place a piece of dry cotton on the site. The patient, with their arm outstretched, holds the cotton in place with their other hand.

13-Remove the needle from the syringe and transfer the collected blood into tubes or vials, with or without anticoagulant, according to the test requested. Slowly drain the blood down the walls of the vials or tubes, without causing foam to form.

14-Cap bottles and tubes to prevent evaporation or contamination. Containers with anticoagulants should be slowly inverted several times, as violent shaking causes hemolysis.

15-Steps 13 and 14 are used when vacuum tubes are not used, where the blood is already conditioned and waiting to be used.

The system uses neutral, reinforced, previously cleaned and siliconized glass tubes, with or without antaculant. The tubes are closed with a siliconized, perforable rubber stopper and contain enough vacuum to aspirate the required volume of blood. The needles, which are sterile and also siliconized, have a sharp bevel with a point at both ends. At the time of collection, the needle is threaded into a special adapter and the smaller end of the needle is inserted into the adapter. The chosen collection tube is inserted into the adapter, without the stopper being punctured by the needle. The larger end of the needle is inserted into the skin after asepsis and crimping. Once the needle has penetrated the skin, the tube is pushed to the bottom of the adapter. As the tube is self-aspirating, you can immediately see that the needle has penetrated the vein, as the blood starts to flow and the tourniquet is immediately removed.

Some rules have been adopted for a good blood collection:

a) The patient must be psychologically prepared;

b) The collection site must be appropriate in order to obtain a sufficient blood sample for the determinations;

c) If the blood is taken with a needle, it is essential that the needle remains inside the punctured vessel. To do this, we must ensure that the collection site is immobilized. In the case of small children, we have to use helpers to restrain the patient. If these people are laypeople, they must first receive the necessary instructions and be psychologically prepared to accept the discomfort that the maneuver causes the child (we have taken this procedure into account even though our study does not extend to children);

d) Blood should flow easily;

e) If tourniquet is used, it should not exceed one minute;

f) After collection, remove the needle before transferring the blood;

g) Use a dry tube with or without anticoagulant when obtaining plasma;

h) After harvesting, apply compressive hemostasis to the site.

BIBLIOGRAPHY

Alberts, Bruce et all. Molecular Biology of The Cell. 3ed. Ed. Garland.
New York- London. 278-287. 1994.

Andrewes, Thomas. Role of respiratory tract proteasis in infectivity of
influenza A virus. J. Infect. Dis. 155(4), 667-672. 1979.

Assad, Young Woo & Reid, Michael. Translational control by influenza
virus. Identification of cis-acting sequences and trans-acting factors which
way segulate selective viral, RNA translation. J. Biol/ Chem. 270(47)
28433-28439. 1971.

Assad, Young Woo. The relative amount of on influenza A virus segment
present in the viral particle is not affected by a reduction in replication of
that segment. J. Gen. Virol. 76(12), 3211-3215. 1973.

Bahl, A.K, Fundamental of Medical Virology. Lea and Febiger,
Philadelphia, 1975.

Barreto, M.L(Org). Epidemiology, services and technologies in health-Rio
Janeiro: Fiocruz.

Becker, W.B. Virus of vertebrates, Baitliere Tindall, Londan 1966.

Buttidield The isolation and cassificaton of tern virus, Influenza virus
A/victoria/1973.J. Hyg Camb, 64:309-320, 1973

Brumback, B.G., et al. Rapid culture for Influenza virus, types A and B, in
96-wess plates clinical and diagnostic virology 4(251-256) 1995.

Castles, T. Et al Bronchiolitis in tropical South India. Am J Dis Child
1444(9): 10261039,1990.

Chaves, J.R.S. et al. Occurrence of Influenza virus in RJ- Arg bras med,
57(2): 104155, 1993.

Cherian T., Bobo, L., et al. Use of PCR-enzime Immunoassay for
Identification of Influenza A virus matrix RNA in clinical samples negative
of cultivable virus. Journal of clinical microbiology, p 623-628, 1994.

Christie, A. B. Nucleotide sequences of influenza A virus RNA segment 7:
a comparison of five isolates. Nucleic Acids Res, 17(7), 2870. 1969.

Claas, E. C. J. et al. Type-specific identification of Influenza viruses A, B
and C by polymerase chain reaction. Journal of virologycal methods 39(1-
13) 1992.

Cox, N., Xu, X. et al. Evolution of hemagglutinin in epidemic variants and
selection of vaccine viruses. Elsevier Science Publishers B.V. 1993.

Cubie, H. et al Detection of Influenza virus in acute bronchiolitis in infants.
Journal of medical virology. 38:283-287, 1992.

Cradock-Watson, et al Rapid diagnosis of respiratory syncytial virus
infection in children by immunofluorescent technique. J. Clin. Path. 24(4):
308-312, 1971.

Doane, F. W., et al. Rapid laboratory diagnosis of orthomyxovirus
infections by electron microscopy. Lancet, 2(7519): 751-753, 1974.

Dolpheide, Mark Alan. Analysis of translation initiation in the influenza B
virus RNA segmente 6 bicistronic gene and posttranslational nodification

of the NB gene product. Dss. Abstr. Int. B, 49(5), 1548. 1980.

Downhan, M.A.P.S et al Infection. Epidemiology and clinical London 1967.

Evans, A. S. Purification, thioredoxin renatuation, and recon stituted activity of the three subunits of the influenza A virus RNA polymerase. Proc. Natl. Acad. Sic. 85(21), 7907-7911. 1076.

Evans, A. S and Olson, B. Rapid diagnostic methods for Influenza virus in chinical specimens, a comparative study. Yale J. Biol. Med., 55(5, 6): 391-403, 1982.

Exaterday, Charles Christorpher. Studies of the immune response to synthetic peptides representing immune determinants of influenza virus hemagglutinin. Diss. Abstr. Int. B 47(7), 2829-2830, 1968.

Felds et al Virology. 150:57-64. 1993.

Gardner, P. S. Review article: Rapid virus diagnosis. J. gen. virol. 36(1) 1-29, 1977.

Gelzen, S, Born et all virology 140:10-16. 1987.

Golthing, M. D. Influenza A virus - specific H in the hemagglutinin HA2 subnit of A/U dorn/72. Virology 214(2), 445-452. 1979.

Goes, Nancy Lugene. Studies on the intracellular and polarized surface tranport of the influenza virus hemagglutinin= a structure/function analusys. Diss. Abst. Inst. B 47(10), 4057-4060. 1954.

Gorman O. T., et al. Evolutionary processes in Influenza viruses: Divergence, Rapid Evolution, and stasis. current topics in microbiology and immunology, vol 176,
1992.

Gruber, A.V. et al. Inhibitory effect of 3'- amino-and 3-deoxyribonucleoside- 5'- phosphatates on RNA- synthesis catalyzed by influenza A viral RNA polymerase and cellular RNA polymerase.Mol. Mikrobiol. Virusuol. 1991,(1), 29-33(Russ).

Hegele, R. G., Robinson, P.J., et al Effects on human the Influenza virus. Am. Rev. Respiratory. Dis; 145: A 434 1992.

Hinshaw,V.S. Isolation and identification of Myxoviruses from domestic and imported avian species, Avian Dis, 17:15-159.1979.

ICN - virology ICN's scientific publication USAFARMA Instituto Farmacéutico N0 02 July 1976.

Informe Epidemiológico do SUS/Centro Nacional de Epidemiologia, coor._Brasília: Fundagao Nacional de Saúde, 1998

Housworth, J. & Langmuir, A. D. Am. J. Epidemol. 100,40-48. 1974.

Kilbourne, M. W. Antiviral immunity induced by recombinant nucleoprotein of influenza A virus, Characteristics and cross-reactitity of T cell responses. J. Immunol. 141(11), 3980-3987, 1975.

Kilbourne, M. W. Acute respiratory infections in infectious diseases in the world. Elsevier Science Publishres 1993.

Jawetz, Ernest et al. Medical Microbiology. Ed. Guanabara koogan. RJ 1991.

Langmuir, A. D. & Houswoth, J. Bull. Worls Health Organ. 41, 393-398. 1969.

Langmuir, A. D & Houswarter, R. G. Evolution of Influenza viruses: rapid evolution and stasis. Mol. Basic virus evol. 531-543. 1969.

Laver, Kibourne. Nuclear localization of all three influenza polymerse proteins and a nuclear signal in polymerase PB2. EMBO J., S(9), 2371-2376, 1966.

Lepene, M. Synergistic role of staphylococal proteases in the induction of influenza virus pathogenicity. Virology, 157(2), 421-430, 1964.

Lilan and French virology 150:60-87, 1988.

Ludwig, S. And Haustein, R. Advanced Laboratory techniques for Influenza Diagnosis. Public Health Service, US Department of Health, Education and

Welfare. Center for Disease Control, Atlant, GA.1992.

Lobo, B.M. Occurrence of respiratory viruses in RJ(Experiencias do Instituto de Microbiologia 1957-1980)- Arg Brás Méd, 57(2): 55-68, 1983

Lui, K.J. and Kendal, A.P. Pathological changes in virus infections of lower respiratory tract. Jr. Clin. Patlol 23(1):7-18. 1985.

Mandel at el virology 220:30-35 1990.

Melo, W.A. Structure and variation Influenza virus, proceeding of International workshop on structure and variation in Infe. Virus. Therebdo, Australia 1977

McDonald, J. C., Miller, D. L., Zuckerman, A. J. & Pereira, M. S. J. Hyg. Biochemical features of the M1 protein matrix in influenza virus C. Mol. Biol. 27(5), 1113- 1125(Cambrigde) 1962.

McDonald, J. C. Siganl processing, glycosulation, and secretion of mutant hemagglutinis of a human influenza virus by saccharomyces cerevisiae. Mol. Cell. Biol, 7(4), 1476-1485, 1962.

Mc Intosh, K. Virology. New York: Raven Press Ltd., 20ed. 411-440p. 1990.

Mintz, L. et al. Nasocomial respiratory virus infections in intensive care nursing: rapid diagnosis by direct immunofluorescence pediatrics, 64(2): 149-153 1996.

Moura, Roberto A. de exames de laboratório, ed Atheneu, SP 1998.

Oliveira, L.H.S, Human Virology. Cultura Médica. RJ. 341p. 1994.

Orstawik, I. et al. Rapid immunofluorescence diagnosis of Influenza virus infections among children in European countries. Lacet. 1:33, 1980.

Palmer, D. F. et al. Advnced laboratory techniques for Influenza diagnosis Public Health Service, US dept of health, education and welfare. Center for disease control, Atlanta, GA 1975.

Pereira, M. S. Gobal surveilhance of Influenza. British Medical Bulletin. vol 35, N0 01, pp 9-14 1979.

Sá, J.P.O. et al, Incidence of Influenza A and B viruses among the population of Maceió-AL,Brazil. Revista Brasileira de Analises Clinicas-RBAC, vol.31(1):9- 11,1999.

Salmidt and Emmons virology 150:70-75, 1989.

Scholtissek C., et al. The role of swine in the origin of pandemic Influenza. Elsevier Science Publishers B.V. 1993.

Shankkinen, H. K., et al. Detection of respiratory virus Influenza by radioimmunoassay and enzime immunoassay on nasopharyngeal specimens from children with acute respiratory diseases. J. clin. Microbiol., 13(2):258-265, 1981.

Siqueira, M.M. et al. Rapid diagnosis of Adenovirus infections: antigens detection in nasopharyngeal secretions. Rev. Bras. Patol. Clin. 22(4): 113-117, 1986.

Stokes, C. E. Epidemiology of acute respiratory infection in population of developing countries. Rev Infect Dis 13(suppl 6/5454-5462, 1988.

Sturdy, P.M., Mcquillin, J. & Gardner, P.S.A comparative study of methods for the diagnosis of respiratory virus infections in childhood. J. Hyg., 67(4): 659670.1969.

Study, P, Daniel and Trs, I. Abba. Basic Immunology.Ed Prentice/Hall do Brasil Rj. 1969.

Tarantino, Bear Denelli. Pulmonary Diseases 80 ed. RJ Ed Guanabara Koogan 1990.

Vdptsa: Influenza activity-wandwide, mach-agoat 1971, Cohem 1997

Watsh, E.E., et al Treatise on Internal Medicine: Virus Influenza, Trad. Amauri J. Da Cruz and others 180 1995.

Ward, Alexander. Prolonged shedding of oman tadine - resistand influenza Aviruses by immunodificiente patients: Detection by polymerase chain resection-restriction analysis. J. Infect. Dis 172(5), 1352-1355, 1995.

Waterson, A. P. Two kinds of myxovirus. Nature(London), 193, 1163. 1962.

WHO. Rapid laboratory techniques for the diagnosis of viral infections. Report a who scientific group. who tchenical report series(661), 1980.

Webster, Jack. Influenza virus- induced encephalopathy in mice: interferon production and natural killer cell activity during acute infection. J. Virol, 60(3), 1062-1067, 1970.

Webster, Jack. Evolutionary patterns of influenza A and C viruses in man. Diss. Abstr. Int. B47(9), 3640-3645, 1973.

Yalken, R.H. Enzime immunoassays for the detection of infectious antigens in body
fluids: current limitations and future prospects. Rev. Infect. Dis. 4(1): 35-68, 1982.

Printed by Books on Demand GmbH, Norderstedt / Germany